Byzantine Military Manuals as Literary Works and Practical Handbooks

This book studies the *Sylloge Tacticorum*, an important tenth-century Byzantine military manual. The text is used as a case study to connect military manuals with the challenges that Byzantium faced in its wars with the Arabs, but also with other aspects of Byzantine society, such as education, politics and conventions in the productions of literary texts and historical narratives.

The book explores when the *Sylloge* was written and by whom. It identifies which passages from classical or earlier works were incorporated in the *Sylloge* and explains the reason why Byzantines imitated works of the past. The book then studies the extent to which the *Sylloge* was original and how innovation and originality were received in Byzantine society. Despite the imitation, the author of the *Sylloge* adapted and updated his material to reflect the current operational needs, as well as the ideological, cultural and religious context of his time. Finally, the book attempts to estimate the extent to which Byzantine generals followed the advice of military manuals and to explore whether historical narratives can be safely used to draw information as to how the Byzantines and the Arabs fought.

Therefore, along with a detailed study of the *Sylloge Tacticorum*, this monograph also addresses broader issues of the pen and the sword, such as military manuals in connection with Byzantine warfare, politics, literature, historiography and education.

Georgios Chatzelis received his PhD from Royal Holloway, University of London in 2017. He has taught at Royal Holloway, University of London as Visiting Teacher. He has delivered guest lectures, papers and communications at international and national conferences, and together with Jonathan Harris he has published an English translation of the *Sylloge Tacticorum*, entitled *A Tenth-Century Byzantine Military Manual: The Sylloge Tacticorum*.

Byzantine Military Manuals as Literary Works and Practical Handbooks

The Case of the Tenth-Century *Sylloge Tacticorum*

Georgios Chatzelis

Routledge
Taylor & Francis Group

LONDON AND NEW YORK

First published 2019
by Routledge

2 Park Square, Milton Park, Abingdon, Oxfordshire OX14 4RN
52 Vanderbilt Avenue, New York, NY 10017

Routledge is an imprint of the Taylor & Francis Group, an informa business

First issued in paperback 2020

British Library Cataloguing-in-Publication Data
A catalogue record for this book is available from the British Library

Library of Congress Cataloging-in-Publication Data
Names: Chatzelis, Georgios, author.
Title: Byzantine military manuals as literary works and practical handbooks /
    Georgios Chatzelis.
Description: First edition. | London ; New York, NY : Routledge, an imprint
    of the Taylor & Francis Group, 2019. | Includes bibliographical references
    and index.
Identifiers: LCCN 2018048178 | ISBN 9781138596016 (hardback : alk. paper) |
    ISBN 9780429488054 (ebook)
Subjects: LCSH: Sylloge tacticorum. English | Military art and science—Early
    works to 1800. | Byzantine Empire. Stratos—Drill and tactics. | Tactics—
    Early works to 1800. | Byzantine Empire—Civilization. | War in literature.
Classification: LCC U43.B9 C47 2019 | DDC 355.009495—dc23
LC record available at https://lccn.loc.gov/2018048178

ISBN: 978-1-138-59601-6 (hbk)
ISBN: 978-0-367-66239-4 (pbk)

Typeset in Times New Roman
by Apex CoVantage, LLC

To all who make us feel significant, from a to Z

Contents

Acknowledgements

This book is based on my PhD thesis, compiled between 2013 and 2017 at Royal Holloway, University of London with funding from the Hellenic Institute. To the director of the latter, Dr Charalambos Dendrinos, I am indebted for his support. My thesis was prepared under the supervision of Prof. Jonathan Harris. It is to the latter whom I owe my deepest gratitude, since his patience, guidance, advice, suggestions and corrections greatly contributed to the drafting of my thesis and proved invaluable to broaden its horizons and approaches. It goes without saying, of course, that any shortcomings or mistakes are solely mine. I would also like to thank the examiners of my PhD thesis, Prof. John Haldon and Dr Catherine Holmes, whose valuable comments, suggestions and ideas during the viva greatly assisted in the production of this book. I am also grateful to the anonymous reviewers, since their careful reading contributed to the shaping of this monograph and yielded numerous corrections and suggestions. My thanks also go to Michael Greenwood of Routledge, whose support facilitated the acceptance and publication of this monograph. For the last stages I owe thanks to Chris Mathews and his associates for their assistance in the production stage. Last but not least, I would like to thank my family, Aristeidis and Lemonia in particular, for all their support, and all my friends and loved ones, who, through the simplest everyday things, make the world a better place.

Abbreviations

BCH: *Bulletin de Correspondance Hellénique*, Paris 1877–.

BF: *Byzantinische Forschungen*, Amsterdam 1966–.

BMGS: *Byzantine and Modern Greek Studies*, Oxford and Birmingham 1975–.

BS: *Byzantinoslavica*, Prague 1929–.

BZ: *Byzantinische Zeitschrift*, Leipzig, Munich and Cologne 1892–.

DAI: Constantine VII, *De Administrando Imperio*, G. Moravcsik (ed.), trans. R.J.H. Jenkins, *Constantine Porphyrogenitus, De administrando imperio*, Washington, DC 1967.

DC: Constantine VII, *De Cerimoniis*, J.J. Reiske (ed.), *Costantini Porphyrogeniti imperatoris de cerimoniis aulae Byzantine libri duo*, Bonn 1829; Eng. Trans. A. Moffatt and M. Tall, *Constantine Porphyrogennetos: The Book of Ceremonies*, Canberra 2012.

DOP: *Dumbarton Oaks Papers*, Washington, DC 1941–.

DRM: *De Re Militari*, In: G.T. Dennis (ed. and trans.), *Three Byzantine Military Treatises*, Washington, DC 1985: 246–327.

DV: Nikephoros II, *De Velitatione*, G. Dagron and H. Mihăescu (eds. and trans.), *Le traité sur la guérilla (De velitatione) de l'empereur Nicéphore Phocas (963–969)*, Paris 1986.

GRBS: *Greek, Roman and Byzantine Studies*, Durham 1958–.

JGR: I. Zepos and P. Zepos (eds.), *Jus Graecoromanum*, Vol. 1–2, Athens 1931.

JÖB: *Jahrbuch der österreichischen Byzantinistik*, Vienna 1969–.

L: Manuscript *Laurentianus Plut.* 75.6, Biblioteca Medicea Laurenziana, Florence.

LD: Leo Deacon, C.B. Hase (ed.), *Leonis Diaconi Caloensis historiae libri decem*, Bonn 1828; Eng. trans. A.M. Talbot and D.F. Sullivan, *The History of Leo the Deacon, Byzantine Military Expansion in the Tenth Century*, Washington, DC 2005.

LT: Leo VI, *Taktika*, G.T. Dennis (ed. and trans.), *The Taktika of Leo VI*, revised edition, Washington, DC 2014.

MS: Maurice, *Strategikon*, G. T. Dennis (ed.), trans. E. Gamillscheg, *Das Strategikon des Maurikios*: Vienna 1981; Eng. trans. G.T. Dennis, *Maurice's Strategikon. Handbook of Byzantine Military Strategy*: Philadelphia 1984.

ODB: *Oxford Dictionary of Byzantium*, Kazhdan, A.P., Talbot, A.-M., Cutler, A., Gregory, T.E. and Ševčenko, N.P. (eds.) 1991. 3 vols. New York and Oxford.

PM: Nikephoros II, *Praecepta Militaria*, In: E. McGeer (ed. and trans.), *Sowing the Dragon's Teeth: Byzantine Warfare in the Tenth Century.* Washington, DC 1995: 12–59.

PS: Syrianos Magister, *Peri Strategias*, In: G.T. Dennis (ed. and trans.), *Three Byzantine Military Treatises:* Washington, DC 1985: 11–135.

REB: *Revue des Études Byzantines*, Paris 1944–.

RSBN: *Rivista di Studi Bizantini e Neoellenici*, Rome 1964–.

SBN: *Studi bizantini e neoellenici*, Rome 1924–1946.

SCE: *Synaxarium Ecclesiae Constantinopoleos*, H. Delehaye (ed.), *Acta Sanctorum* 62, Brussels 1902.

ST: *Sylloge Tacticorum*, A. Dain (ed.), *Sylloge Tacticorum quae olim 'Inedita Leonis tactica' dicebatur*, Paris 1938.

TC: Constantine VII, Theophanes Continuatus, In: I. Bekker (ed.), *Theophanes Continuatus, Ioannes Cameniata, Symeon Magister, Georgius Monachus*, Bonn 1838: 3–481.

TC   (b): Constantine VII, Theophanes Continuatus, J.M. Featherstone and J.S. Codoñer (ed. and trans.), *Chronographiae quae Theophanis Continuati nomine fertur libri i–iv*, Boston and Berlin 2015.

TIB: *Tabula Imperii Byzantini*, Belke, K., Hellenkemper, H., Hild, F., Koder, J., Küzler, A., Merisch, N., Restle, M., Soustal, P., Todt, K.-P. and Vest, B.A.1976–2014. 15 vols. Vienna.

TM: *Travaux et Mémoires*, Paris 1965–.

TNO: Nikephoros Ouranos, *Taktika*, In: E. McGeer (ed. and trans.), *Sowing the Dragon's Teeth: Byzantine Warfare in the Tenth Century.* Washington, DC 1995: 88–163.

VB: Constantine VII, *Vita Basilii*, I. Ševčenko (ed. and trans.), *Chronographiae quae Theophanis Continuati nomine fertur liber quo Vita Basilii imperatoris amplectitur*, Berlin 2011.

ZRVI: Zbornik Radova Vizantoloshkog Instituta, Belgrade 1952–.

Introduction

What is the *Sylloge Tacticorum*?

The *Sylloge Tacticorum*, or *Συλλογή Τακτικῶν* (hereafter *ST*), which translates to *Compilation of Tactics*, is a tenth-century Byzantine military manual attributed to Emperor Leo VI (886–912), bearing the date 903–904 AD in its title. However, since this attribution is generally not accepted by modern scholarship, a more specific dating remains open to question. The *ST* belongs to the literary genre of military manuals which stretches back to antiquity. The first military manual appeared in the fourth century BC; it was the work of Aeneas the Tactician who wrote a treatise on how to withstand a siege.[1] Such works continued to be produced throughout antiquity, and by the late sixth or early seventh century the most influential Byzantine military treatise appeared under the name *Strategikon*, attributed to Emperor Maurice (582–602) (hereafter *MS*). The next extant military manual, the *Peri Strategias* (hereafter *PS*), written by Syrianos Magister, seems to date sometime in the ninth century.

The tenth century was an important milestone for the genre of military treatises. The general context of high literary production under the so-called 'Macedonian renaissance', in combination with the Arab threat, facilitated the production of several military manuals. One of these was the *ST*, which was strongly influenced by its predecessor, the *Taktika* of Emperor Leo VI (hereafter *LT*). In turn, the *ST* heavily influenced its successor, the *Praecepta Militaria* (hereafter *PM*), which is attributed to Emperor Nikephoros II Phokas (963–969).

The *ST* consists of 102 extant chapters, which can be divided into three main categories. The first category covers chapters 1–56, where a wide variety of military matters is discussed, including generalship, sieges, battle and marching formations, division of booty, raids, ambushes, encampment, posting of officers, spies and truces. The second category includes chapters 57–75, which contain information regarding war by other means, discussing matters such as protecting against poisonous food or drinks, poisonous arrows, using flammable mixtures or easily neutralizing enemy horses. The third and final category covers chapters 76–102, all of which contain stratagems and anecdotes of military commanders in ancient times.

Manuscript tradition

The *ST* is currently reported to be preserved in three manuscripts. The oldest available manuscript is the *Laurentianus Plut.* 75.6 (hereafter L), whereas the other two are more recent copies: the *Bernensis* 97, which is a direct copy of L, is dated to the sixteenth century, and the *Parisinus* 2446, which is a direct copy of the *Bernensis* and is dated to the seventeenth century.[2] L most probably dates to the fourteenth or early fifteenth century, and it is a codex measuring 190 by 270 mm, consisting of 278 folios in total.[3]

The earliest recorded history of L can be traced to 1491, when the codex is found in the list of books which Janus Laskaris bought from Corfu. Janus also visited the Avramis library while in Corfu, but L does not seem to have been part of this library, since it is not featured in Janus' book list, which specifically listed items coming for the Avramis library.[4] Janus, who was a prominent Greek scholar of his time, was patronized by the Medici, perhaps with the aid of Demetrius Chalkokondyles, sometime after the death of Bessarion (1472). In the context of his activities in Florence as a Greek scholar, Janus undertook two visits to Greece in search of Greek manuscripts. In his voyage between 1491 and 1492 Laskaris visited Corfu, where he purchased our codex, among other works.[5] The manuscript, however, did not directly go to the patron of Janus, Lorenzo de' Medici, as it does not feature as a holding of the Medici library in 1495. Nevertheless, we can trace its existence in the possession of the Medici to 1508, as the manuscript appears in the inventory of Vigili, listed as number 381, and in folio 108v of the so-called list of Hannover, which is a list of works and authors featuring the books that Laskaris introduced to Florence from his travels.[6] The L is still in the Medicea Laurenziana library in Florence today.

The codex contains various works of different kinds. More specifically folios 1–71 contain medical treatises, some of which are dedicated to Constantine Porphyrogennetos.[7] Folios 116–124 preserve a series of military laws and military hymns, which both have been edited and studied by modern scholars.[8] After that, folios 124–247 contain the *Hippiatrika*, a veterinary treatise, which is most probably linked to Constantine VII (945–959).[9] The remaining folios (247–275) are filled with various treatises, mostly small in size, some of which are dedicated to medicine or geography.[10]

The *ST* occupies folios 72r–116r. A *pinax*, namely a table of contents, with the title of each chapter, covers folios 72r–73v. However, the *pinax* continues over to folios 73v–74v to cover not only the contents of our treatise but also the contents of the other two works that follow the *ST* in the codex (i.e. the *Poinalios Stratiotikos Nomos* and the *Akolouthia*). This implies that the *pinax* was not initially part of the *ST* but was added at a later date. Alphonse Dain argued that it was not added by the copyists of L, as some light corrections strongly imply a more ancient origin. Most probably the *pinax* was added sometime after the creation of the *ST*, when the latter was included in a corpus dedicated to military matters that contained all three works, and thus served as a general table of contents for the whole codex.[11] The next folios (74v–103v) preserve chapters 1–75. The last part occupies folios 103v–116r, at the beginning of which, a new title appears,[12] and directly after it follow the chapters 76–102.[13]

According to the author of our treatise, the *ST* was originally accompanied by two diagrams which unfortunately are not preserved in any of our surviving manuscripts. Both were dedicated to infantry formations.[14] The first diagram seems to have depicted a battle array consisting exclusively of infantry, while the second probably illustrated a battle formation in which the infantry was drawn up in a square and the cavalry was stationed inside it. A small, rough sketch of this square formation is preserved into the right lower margin of L, folio 95r. This sketch, however, bears little resemblance to other surviving diagrams of military treatises, and it seems, therefore, that it was not a copy of the original one, but rather a later design. This sketch could have been the work of the copyists or of an older owner of the text and could have served to facilitate a better understanding of the guidelines of the text.[15]

The *ST* was copied by two different copyists. Daniele Bianconi has recently argued that one of them was Krateros, a scribe who belonged to the milieu of the fourteenth-century scholar Nikephoros Gregoras. The environment of Gregoras seems to have played an active role in the drafting of L in general, since yet another of his scribes, known as Anonymous G., seems to have collaborated in the writing of the codex.[16] The first hand copied from the beginning of the *ST* and up to the end of chapter 67, while the second hand continued his work up to the end.[17]

Although the second title of the treatise states that it will cover the deeds of ancient men in twenty-eight chapters, only twenty-seven chapters appear in the treatise.[18] Furthermore, L does not preserve chapters 68–73 and half of chapter 74. The first scribe comments on this loss in the margins of the manuscript, reporting that the chapters were already missing from the manuscript he used. The copyist attempted to fill in some of the missing chapters using a lost source, but although the *pinax* preserved the original titles, the missing chapters were filled in the wrong order. To make matters worse, one of the chapters that was filled in by the scribe does not correspond with any of the original material of the *ST*, as its text is irrelevant to the chapter headings that are found in the *pinax*.[19]

Dain explained this inconsistency by trying to identify the method of the copyist. He suggested that the copyist first attempted to complete chapter 74 since it was already partly preserved. In order to accomplish this, our copyist used a lost source, which Dain named *Corpus Perditum*. According to Dain, the copyist managed to fill in chapter 74 by drawing on chapter 38 of the *Corpus Perditum*, but as this was the first chapter which he attempted to fill, the copyist numbered it 68, probably because this was the first missing chapter of the *ST*. Consequently, he added it in the margins of folio 103r, which was the first margin available. His next step was to fill the next missing chapter he could find successively by reading the lost source in reverse order, from the end to the beginning. However, despite the fact that chapters 71–73 of the *ST* correspond with chapters that the *Corpus Perditum* probably preserved, our scribe either failed to notice their existence, or perhaps had a version which was mutilated.[20]

In any case, the next chapter of the *ST* which the copyist managed to locate in the *Corpus Perditum* was chapter 70. According to Dain this corresponds with chapter 31 of the lost source. The scribe added this chapter in the left margin of folio 102v, namely the one right before folio 103r, as he was aware that in the order of the *pinax*, this chapter came first. Nevertheless, he numbered it chapter 69, as it was the second in line which he added. As he continued the reading of *Corpus Perditum* in the same reverse order, he came across chapters 29–27, some of which correspond with missing chapters of the *ST*. Consequently, our scribe turned to the next available folios 103v–104r and filled the missing chapter 69 with chapter 29 of the *Corpus Perditum*. Once again, as this was his third addition, he numbered it chapter 70. Nevertheless, as the next missing chapters were preserved one after the other in the *Corpus Perditum*, the scribe accidentally copied chapter 28 of the lost source, despite the fact that its text is irrelevant to the original contents of the *ST*. Be that as it may, he numbered this chapter 71, since it was his fourth addition. The next chapter of the *Corpus Perditum* in line was chapter 27, which corresponds with the surviving chapter 65 of the *ST*, so our copyist correctly skipped it. Finally, he filled chapter 68 of the *ST* with the next chapter of the lost source in line, chapter 26, and since this was his fifth addition, he numbered it chapter 72.[21]

Editions and translations of the *Sylloge Tacticorum*

In 1770, Angelo-Maria Bandini was the first scholar to re-discover the existence of the *ST* in L, as he was preparing his catalogue of the manuscripts of the Medicea Laurenziana library.[22] The first edition of the text, however, was only published 84 years later by Hermann Köchly.[23] His work appeared in 1854; it consisted of two volumes which contained, among other works, chapters 31–33, 38–39, 41–43 and 53–55 of the *ST*, which were edited using the *Bernensis* 97 rather than L.[24] In 1863, Jacques-Paul Migne included Köchly's text in the *Patrologia Graeca*, where a facing Latin translation was also added.[25]

In 1887, twenty-four years later, Johannes Melber, who had a strong interest in the work of Polyaenus, edited the relevant chapters 76–102 of the *ST*, based on L.[26] In 1917, Rudolph Vári included in his edition of *LT* some previously unedited chapters of the *ST*, as a sort of a critical apparatus, in order to compare and contrast the information given.[27] The last partial edition appeared in 1932, when Jean-René Vieillefond, who was interested in the tradition of the *Cesti* of Julius Africanus, edited chapters 57–75, along with a useful study regarding the sources and the dating of the *ST*.[28]

The first complete edition of the *ST* was published in 1938 by Dain, who based his edition on L. Dain's edition included all of the chapters of the manual, as well as the chapters that were added by the copyists in the margins of the manuscript. The edition of Dain remains the only complete one. In 1939, Dain included a French translation of the five chapters which were filled in by the copyist of L.[29] In 1994, Everett Wheeler published the first partial English translation of the *ST*, which included chapters 76–102. Wheeler based his translation on Melber's text and not that of Dain, but an appendix with the different readings was included in his book. A new English translation of the complete text, along with some corrections of Dain's readings, as well as with a critical introduction and notes, was published by Jonathan Harris and Georgios Chatzelis in 2017.

Previous analysis of the *Sylloge Tacticorum*

When, in 1770, Bandini included the *ST* in his catalogue, he described some of its contents and pointed out that although the title of the treatise is attributed to Leo VI, its contents have many differences compared to *LT*.[30] The next century saw a number of scholars dedicating some of their focus to the *ST*, the first of whom was Friedrich Haase. In 1847, Haase commented on the importance of the *ST*, stating that the treatise is at least of equal value and deserves at least the same attention as *LT*. Haase argued that the *ST* pre-dated *LT*, but he saw it as an unedited treatise of Leo VI.[31] The phrase *Inedita Tactica Leonis* was thus introduced and accompanied the *ST* for at least a century to come.

Seven years later, Köchly was a little more cautious on the dating issue. He argued that it is not certain whether *LT* pre-dated the *ST* or vice versa, but he did not question the attribution and authorship of both works to Leo VI.[32] Köchly also attempted to identify the sources of the *ST*. In the introduction of his partial edition he commented that part of the treatise derived from Aelian, *MS*, the *PS* and *LT*. In addition, he was the first scholar to point out that L is the oldest available manuscript of the *ST* and that the *Bernensis* 97 is only a direct copy of it.[33]

The picture remained unchanged throughout the nineteenth century. Therefore, in 1863 Migne included a reprint of Köchly's edition of the *ST* in the *Patrologia*, under the name of Leo VI. Likewise, in his history of Byzantine literature, published in 1887, Karl Krumbacher treated the *ST* as an addition to *LT*.[34] It was from the beginning of the twentieth century that the study of the *ST* intensified. At that time, the most influential studies on the treatise were published which managed to direct scholars to the right direction.

In 1927, Vári produced the first study solely dedicated to the *ST*. In his article, he discussed various aspects of the text. He agreed that L was the principal manuscript and explained that the *Parisinus* 2446 was a direct copy of the *Bernensis* 97. Apart from these, however, Vári was the first to argue that, in spite of the title of the treatise, Leo VI was not the real author of the *ST*. He underlined that there is a difference in style and sources between *LT* and the *ST* and looked to internal evidence to find an alternative explanation. Vári stated that the phrases 'our Majesty' and 'our generals' were proof that the manual was indeed written by an emperor and that the identical military hierarchy in both manuals, as well as the mutual appearance of the statement that the hand-siphons were recently invented, indicated that the *ST* was written in the time of Leo VI.[35] Vári therefore proposed that the manual was originally written by Alexander, the brother of Leo VI, who probably completed his work before *LT*. After Leo VI became increasingly suspicious of his brother, however, he arrogated the treatise for himself, and once *LT* was completed, he reduced the *ST* to a mere reference work.[36]

Despite the fact that Vári's theory was never accepted by modern scholarship, the importance of his contribution cannot be overlooked. He was the first scholar to break the connection between the *ST* and Leo VI after at least a hundred years. Although the attribution of the *ST* to Alexander lacked cogency, Vári's article was the first and last detailed study dedicated to tackling the problem of dating and attribution of the *ST*. All subsequent views on dating and attribution were presented without detailed argument and analysis, as they were mostly expressed in passing.

In 1932, Vieillefond provided two fundamental insights on the issue of the sources of the *ST*, which were to become the basis of modern scholarship. First, he recognized that chapters 57–75 originally derived from a lost source, which he named *Corpus X*. Vieillefond identified that the extant source which preserves the most similar phrasing to that of chapters 57–75, is, in fact, the *Apparatus Bellicus*, and not the *Cesti* of Julian Africanus, as Vári had previously argued.[37] With this evidence in mind, Vieillefond concluded that the *ST* dated in the late tenth century or in the beginning of the eleventh.[38]

Vieillefond's argument regarding the lost sources of the *ST* was further advanced by Dain in 1938–1939. After Dain had made a comparative study of the *Apparatus Bellicus*, *ST* and *Taktika* of Nikephoros Ouranos (hereafter *TNO*), he concluded that it was not only chapters 57–75 which derived from this lost source, but the whole second half of the treatise. He therefore proposed that chapters 56–102 derived from what he preferred to call *Corpus Perditum*.[39] Accordingly, Dain assumed that another lost source was used for the first half of the manual, a source which he called *Tactica Perdita*. He supported this theory by explaining that although in chapters 21.4 and 22.5, the author of the *ST* writes 'as I have already said', his cross-references are, in fact, inaccurate as no relevant information had been provided earlier.[40] Dain consequently concluded that the inaccurate cross-references, as well as the phrase 'our Majesty', which only appears in the first half of the treatise, were accidentally copied in the *ST* from the *Tactica Perdita*. Whatever the case, Dain identified most of the traceable sources that our author had used. He noted that the extant texts which have a similar phrasing to that of the *ST* are Aelian, Julian of Ascalon, Onasander, *MS*, *LT*, the *PS*, the *Apparatus Bellicus*, the *Hypothesis* and the *Ekloga*.[41]

Given that Dain saw the author of the *ST* as a careless copyist, he wondered whether the *ST* really was a manual produced for current needs. Although he recognized that some parts of the treatise were original, he expressed the view that the manual should be considered either partly or entirely bookish, a literary work, a work of the library.[42] Finally, Dain argued that the manual dated from the middle of the tenth century without, however, going into more detail.[43] Around thirty years later, he confessed that a dating at the time of Leon VI was more probable, once again without further specifying.[44] Although Dain changed his mind on the dating, it was his initial theory that influenced modern scholarship the most. The majority of the scholars who referred to the *ST* commented that its date remained uncertain but most probably dates around 950.

In 1978, Herbert Hunger dedicated some pages of his survey of Byzantine literature to the *ST*. Hunger more or less repeated the findings of Dain. He emphasized that the author of the *ST* did not exhibit the skills of a successful editor and explained that some parts of his work should be attributed to a direct copy from lost sources even if they are in the first person. Hunger further commented that the poor editing skills of our author was the reason why some of the material is treated both in the first and in the second half of the treatise.[45]

Between 1991 and 1995, Eric McGeer was one of the last scholars who studied the *ST* in some detail. McGeer accepted Dain's initial dating theory and went it a step further. He argued that the fact that L includes medical and veterinary treatises dedicated to Constantine VII should imply that the *ST* was also compiled during his reign. In addition, McGeer noticed that the earliest testimony for the appearance of the Byzantine heavy cavalry, the *kataphraktoi*, is the battle of al-Hadath (954). With that in mind, he proposed that since the *ST* refers to the *kataphraktoi*, who also appear in the *PM*, it should have dated around this period, and more specifically around 950. McGeer's contribution, however, was not only confined to the issue of dating. He also noted that the *PM* used the *ST* as source and underlined some important innovations and evolutions as they appear in the two manuals, the most important of which are the hollow-square formation, the tactics of the *menavlatoi*, the three-lined cavalry formation and the tactics of the *kataphraktoi* and the *prokoursatores*.[46]

In his monograph on Nikephoros II Phokas, published in 1993, Taxiarchis Kolias took a different approach on the dating issue. He refuted the idea that the *ST* dates to the reign of Constantine VII and argued instead that the material of the treatise better fits the context of the first half of the tenth century. According to Kolias, the fact that the *ST* is the first manual which refers to the *kataphraktoi* cannot be convincingly used to support a dating around 950. He argued that the information of the text is in accordance with the first half of the tenth century and that the *kataphraktoi* might have appeared at that time. Unfortunately, Kolias did not elaborate further on the matter. He only noted that if the *ST* dates around 950, we must not consider it as a source which provides absolutely contemporary information.[47]

A year later, in 1994, Everett Wheeler discussed the *ST* as part of his work on Polyaenus. In his introduction, he argued that the author of the *ST* had manipulated the tradition of Polyaenus, since chapters 77–102 preserved a different version of stratagems compared to our most relevant extant source, the *Hypothesis*.[48] More specifically, he stressed that the author of the *ST* included new stratagems and information, while others are listed in different order and with different phrasing. Wheeler took Dain's argument, that the *ST* is a purely literary work, a step further, arguing that the fake attribution to Leo VI and the inaccuracy of cross-references indicate that the whole treatise is a forgery.[49]

In 2003, Marcel Meulder also turned his attention to the part of the *ST* which preserves the stratagems of classical commanders. Meulder tried to explore the identity of King Merops, who is one of the generals that appears in the *ST* but not in the *Hypothesis*. Meulder discussed the possible sources that these unique passages could derive from, not excluding the possibility of reliance on Persian material as well. He wondered whether Merops was merely a nickname, used to substitute the name of generals who had been forgotten or lost, but from a comparative study of different passages Meulder concluded that Merops could have been the Sassanian ruler Shapur I (240–270).[50]

Six years later, in 2009, Edward Luttwak dedicated a page of his book to the *ST*. Luttwak mostly summarized the views of McGeer and Dain, and agreed with McGeer that the treatise should not only be regarded as a literary work.[51]

The same year, Laura Mecella published an essay on the sources of the *ST*, which is the first work, after that of Dain, to provide some original argument on this issue. Although Mecella supported Dain's view on the existence of the *Corpus Perditum* with additional evidence, she rejected the existence of the *Tactica Perdita*. Mecella explained that Dain's simplified theory that the author of the *ST* used the *Tactica Perdita* for the first half of the treatise and the *Corpus Perditum* for the second is problematic. She showed that this division cannot be accepted, since traces of the *Apparatus Bellicus* and the *PS*, which were both supposedly parts of *Corpus Perditum*, also appear in the first half of the *ST*. Mecella proposed, instead, that the author of the *ST* had only used one major lost source, the *Corpus Perditum*, and that most of the material of the first half of the work was drawn from other individual sources. Mecella closed her essay by commenting that a more detailed study of the sources is necessary to draw more firm conclusions.[52]

A year later, in 2010, Dennis Sullivan included the *ST* in his study of Byzantine military treatises. While he summarized the views of Dain and McGeer, he also noted that the *ST* is the first manual to record new technical vocabulary, such as the word *kompothelykion* to denote loops and buttons for the attachment of sleeves, and the word *mosynas* to describe siege towers. Sullivan also commented on the fact that the author of the *ST* claims to provide the reader with contemporary material, a promise which is fulfilled in chapters 38–39 and 46–47.[53]

A recent work to discuss the *ST* is that of John Haldon. In 2014, as part of his study of *LT*, Haldon provided a small summary of the views of Vári, Dain, McGeer and Mecella. Apart from that, he also addressed the basic issues of the *ST* that remain unresolved. Haldon started to question matters of dating and tradition, wondering whether the text was indeed written in the time of Leo VI but was later revised with the addition of material that seems to date in the 950s, like the *menavlatoi* and *kataphraktoi*. Finally, Haldon also provided some original discussion on the similarities of some tactics from *LT*, to the *ST* and the *PM*.[54]

Finally, in 2017, through his study of the reception of Aeneas the Tactician in Byzantium, Philip Rance contributed to the study of the sources of *ST*, noting that contrary to previous views, it was only two chapters of the *ST* (70 and 76) which originally derived from Aeneas the Tactician. Rance also enhanced our understanding of the *Apparatus Bellicus* and *Corpus Perditum* and agreed with Dain and Mecella that the author of the *ST* paraphrased his sources quite freely.[55]

The purpose of this book

Despite the earlier studies, the *ST* remains a treatise that is vastly understudied. Crucial aspects of the text remain unanswered, some of which are the dating and attribution of the text, the relation of the *ST* to other military manuals; the extent to which the manual was relevant to contemporary challenges of Byzantine warfare; and its connection with the broader literary, political and social context.

The purpose of this book is to provide a fresh insight into the *ST* and to cover this gap. The aim is not only to engage with views of previous scholarship and to provide original argument but also to connect the *ST*, and military manuals in general, with broader themes and aspects of Byzantine studies. There have been many noteworthy works on Byzantine imitation and innovation, as well as on inter-textual analysis. Other significant advances have been made in identifying lost sources and trends in historiography, as well as the purpose of compilation literature in Byzantium. It is therefore important to build on these and to connect the *ST* with them, so as not only to produce research that will be confined to the development of Byzantine warfare, but a book which can be used as a case study to address much broader issues of the pen and the sword, that is, Byzantine warfare in connection with court education, Byzantine literature and historiography.

In this light, Chapter 1 will focus on the context of Byzantine warfare in the period 900–950 in order to demonstrate the challenges and changes of Byzantine strategy in the time that the *ST* was compiled. This will serve as a background chapter which will set the stage to understand the treatise and its contents better.

Chapter 2 will examine the issue of Byzantine Classical imitation and its relation to court education. After this is explained, the study will specifically discuss the sources that the author of the *ST* used. The chapter will list possible sources, but the main aim is to build upon Mecella's study and to determine whether the *Tactica Perdita* ever existed. This will allow us to estimate whether it is safe to regard the text of the *ST* as trustworthy evidence to determine the author and dating of the text, or whether it should be seen as an unsafe testimony which was the result of slavish copying from lost sources.

The next chapter, Chapter 3, will discuss the dating of the manual. After analysing the problems of the dating theories of Vári, Vieillefond, Dain, Kolias, McGeer and Haldon, the study will aim to tackle the issue from a fresh perspective. To achieve this, the internal evidence of the *ST* will be studied comparatively with other sources which shed light on the political, administrative and military milieu of the empire and its neighbours in the first and second half of the tenth century.

Chapter 4 will attempt to explain authorship and attribution. It will take into consideration the views of Vári, McGeer and Haldon and discuss their problems. The issue will be approached with a fresh look at the internal evidence of the text and in accordance with the findings of the new dating theory. The goal will be to identify the original author and the existence of possible ghost authors or redactors.

After the basic issues of sources, dating and attribution are sufficiently discussed, the study will turn into evaluating how up-to-date and original the information of the *ST* was. Chapter 5 will therefore begin by broadly examining tradition and originality in Byzantine literature and how relevant scholarly views have evolved over time. The chapter will then focus on identifying innovations and originality in the *ST*. The material of the treatise will be studied comparatively with its sources to highlight any adaptations or shifts in attitude, but also in comparison with tenth-century Byzantine strategic and tactical needs, as well as with perceptions of warfare, religion and morality. Finally, the original material of the *ST* will be studied comparatively with that of anterior and posterior manuals in an attempt to highlight the gradual evolution that took place over the decades and to determine the place that the *ST* holds in the development of Byzantine warfare.

Chapter 6 will study the purpose of the *ST* and will attempt to determine its practical use. First there will be a presentation of various views concerning the practicality of Byzantine military manuals and compilation literature in general. Second, an analysis of the various problems of these theories will be provided, along with a study of how the Byzantines saw and used their past. In addition, we will examine the role of historiography and literature in providing examples to imitate. After all these matters are explained, the discussion will focus on determining how practical certain aspects of the *ST* were by studying its contents in relation to the testimony of Byzantine, Arab and Western historical narratives, as well as administrative documents and other manuals. The study will aim to look at the sources critically and determine whether they can be taken at face value.

The final chapter, Chapter 7, will identify the impact that the *ST* had on posterior manuals and how its tactics were adapted and evolved. It will also provide some information regarding its influence on Byzantines and Westerners in the eleventh century and beyond.

Notes

1 For an extensive overview of the genre in antiquity from several perspectives, see Rance 2017b: 9–53.
2 Köchly 1854; Vári 1927: 241–2; Dain 1938: 9–10; 1940b: 36; Dain and Foucault 1967: 338; Andrist 2007: 126–37; Wallraff et al. 2012: xl.
3 Bandini 1770: 151; Köchly 1854; Vári 1927: 242; Vieillefond 1932: xlvi–xlix; Dain 1938: 9–10, 1940b: 38–9; Dain and Foucault 1967: 388; Fryde 1996: 425, 612; McCabe 2007: 33, n.101.
4 Müller 1884: 379–407; Jackson 2003: 137.
5 Knös 1945: 25–9, 33–7; Tsagas 1993: 13–19; Speake 1993: 325–30; Harris 1995: 101, 123–4; Irigoin 1997: 485–91.
6 Vogel 1854: 156–7; Fryde 1983: 160–1, 203; 1996: 651; Jackson 1998: 91–3, 101; 2003: 137–9; Markesinis 2000: 302–6; McCabe 2007: 34.
7 For more information on medical treatises connected to Constantine Porphyrogennetos see: Cohn 1900: 154–8; Dain 1953: 70; Hunger 1978: ii.305–6; Lemerle 1971: 296; Sonderkamp 1984: 29–41.
8 For the military laws of Rufus, see: Korzenszky 1931; *JGR*, ii.ix–x, 80–9, and for the military hymns: Pertusi 1948: 145–68.
9 The *Hippiatrika* has been studied in detail by McCabe 2007. For other works on the *Hippiatrika* and Constantine VII see: Cohn 1900: 158–60; Hunger 1978: ii.306; Lemerle 1971: 296–7.
10 Bandini 1770: 150–1; Dain and Foucault 1967: 338; Fryde 1996: 424, 612.
11 Dain 1939: 11–12, n.1.
12 *Strategic Recommendations from the Deeds and Stratagems of Ancient Men, Romans, Greeks and Others, In Twenty-Eight Chapters.*
13 Bandini 1770: 147–9; Dain and Foucault 1967: 338.
14 *ST*, 45.20, 47.8.

15 See, for example, the appendix of diagrams in the edition of *MS* and the *Syntaxis Armatorum Quadrata*.
16 Bianconi 2008: 372, n.104; 2011: 125, n.40; 2012: 311. For the Anonymous G. see: Peréz-Martín 2008: 431–58.
17 Bandini 1770: 148–9; Dain 1939: 12; Dain and Foucault 1967: 338; Mecella 2009: 107.
18 Krentz and Wheeler 1994: xxi.
19 Vieillefond 1932: lii–liv; Dain 1938: 113–15; 1939: 12–14; Dain and Foucault 1967: 353; Mecella 2009: 107–8.
20 Dain 1939: 28–31.
21 Dain 1939: 28–31.
22 Bandini 1770: 148–9.
23 Köchly 1854; Dain 1938: 9.
24 Köchly 1854; Vári 1927: 241; Dain 1938: 9.
25 *Patrologia Graeca*, 1095–120.
26 Melber 1887; Dain 1938: 9.
27 Vári 1917–1922.
28 Vieillefond 1932.
29 Dain 1939: 34–6.
30 Bandini 1770: 148–9.
31 Haase 1847: 17.
32 Köchly and Rüstow 1855: 11.
33 Köchly 1854: 4.
34 Krumbacher 1897: 637
35 τῆς βασιλείας ἡμων; οἱ ἡμέτεροι στρατηγοί
36 Vári 1927: 241–3, 265–70.
37 Vieillefond 1932: lii–liv, c.f. Vári 1927: 265–6.
38 Vieillefond 1932: xlvi–xlvii.
39 Dain 1938: 8; 1939: 14–31, 70–1; Dain and Foucault 1967: 353.
40 ὥς μοι λέλεκται; ὡς ἤδη ἐρρέθη μοι.
41 Dain 1938: 8; Dain and Foucault 1967: 350–1.
42 Dain and Foucault 1967: 351.
43 Dain 1938: 8.
44 Dain and Foucault 1967: 357.
45 Hunger 1978: ii.333.
46 *ODB*: iii.1980; McGeer 1995a.
47 Kolias 1993a: 24–6, n.10.
48 A Byzantine treatise consisting of excerpts of Polyaenus, see: Dain 1937: 73–86; Dain and Foucault 1967: 337; Schindler 1973: 205–16; Krentz and Wheeler 1994: xx–xiii; Wheeler 2013: 53.
49 Krentz and Wheeler 1994: xxi–xxii.
50 Joseph Genesios, 4.34; Meulder 2003: 445–66.
51 Luttwak 2009: 312.
52 Mecella 2009: 100–1, 107–13.
53 Sullivan 2010a: 155.
54 Haldon 2014: 66–8, 337–8, 359, 360.
55 Rance 2017a: 324–58.

1 The context of the *Sylloge Tacticorum*

Byzantine warfare c. 900–950

The *ST* belongs to a particular moment in Byzantine military history: the point when, after years of endemic warfare with the Arabs along a fixed frontier in Asia Minor, the balance began to shift in favour of the Byzantines. Likewise, in the Balkans, the Byzantines had been at war on and off with the Bulgars for several centuries. While they came close to complete defeat at the hands of Khan Symeon in the first twenty years of the tenth century, the final decades of the century saw them going over to the offensive, culminating in the conquest of Bulgaria in 1018.

The status quo on the Arab–Byzantine frontier

On the eve of the tenth century, the Byzantine army and the type of warfare that it waged had crystallized. The evolutions and responses that the rapid Arab conquests had necessitated had become established and consolidated, and warfare had taken on a standard form. From 720 to the third decade of the tenth century, the eastern frontier was more or less unchanged. Both Byzantines and Arabs turned to defence, fortification and consolidation, while military campaigns took the form of raids which did not aim at annexations, only at temporary occupations and plunder.[1]

Against this background, the Byzantine army evolved in order to respond better to the Arab threat. A series of reforms had started to take place slowly and gradually from the seventh century.[2] By the tenth century, the provinces of the empire were organized into *themata*, and each was governed by a *strategos* who had supreme political and military authority in his area. Among the responsibilities of the latter was to supervise the upkeep, recruitment and training of the *stratiotai*, who formed the army of the *themata*. The *stratiotai* were part-time cavalry soldiers, but also local land-owners who held military land. Consequently, they were burdened with military service which was to be fulfilled either by themselves, by the participation of another soldier on their behalf or by paying a sum to the state to hire a mercenary in their place. The *stratiotai* were responsible for buying and maintaining their own arms and armour, as well as horses. The *themata* also provided infantry, but information regarding thematic infantry soldiers is very scarce.[3]

As a political counterweight to the *stratiotai* and the *strategoi*, but also for their supplement and enhancement, the Byzantine Empire had at its disposal professional full-time soldiers, the *tagmata*, who were based in or around Constantinople and were under the direct control of central authority, independent of provincial elites and local strong families. The *tagmata* were created during the reign of Constantine V (741–775), and by the middle of the tenth century there were four of them: the *Scholai*, the *Exkoubitores*, the *Vigla* and the *Hikanatoi*. They constituted the elite of the army and were provided with very satisfactory payments, rewards and donatives. Moreover, their armament, horses and maintenance were the responsibility of the state.[4] However, the nature of warfare was such that the *tagmata* were mainly used for offensive campaigns and deep raids. They were seldom deployed for defence: only when there was extreme pressure by invading armies. Therefore, the frequent task of defending against enemy raiders was mainly undertaken by part-time local soldiers of the *themata*.[5]

As warfare took on established form, the challenges that the Byzantine army faced, from the eighth century up to the beginning of the tenth, were the same. The Arabs conducted raids into Byzantine territory yearly, sometimes even three times in one year. A tenth-century Arabic source, Qudāma ibn Ja'far, describes the established pattern of the yearly raids:

> The most challenging raid of the skilled frontier-raids is the so-called spring time raid, which begins around ten of May after the people have pastured their animals and when their horses are in good condition. This lasts thirty days through the remainder of May and ten days into June. In the land of the Romans, they find pastures for their livestock like a second spring. Then they return home and settle for twenty-five days, being the remainder of June and five days into July until [the beasts are] rested and fattened. Then the people join in the summer raid, starting from ten of July and lasting sixty days. As for winter raids, everyone I know says that if these are necessary, they should not penetrate too far, but only to the extent of about twenty days round-trip, just long enough that the raider can load sufficient provisions on horseback. They should be done by the end of February and they should head out in the early days of March, for at that time they will find the enemy – themselves and their animals – weaker and their livestock greater. Then they return and vie with one another to put their animals to green pasture.[6]

These campaigns, even when they reached deep into Anatolia, were part of a war of attrition and were aimed at extracting booty, destroying the enemy's potential and capturing prisoners and livestock. Sometimes, they also had a punitive character, as they were launched in order to compensate for an annexation or a significant Arab defeat. When fortresses and cities were conquered, they were looted, destroyed and abandoned; the Arabs showed no intention of annexing more Byzantine territory. The conclusion of such raids was usually a short-term truce and an exchange of prisoners.[7]

The Byzantine response to these challenges was shaped during the eighth century and remained almost identical up to the end of the first decade of the tenth. The Byzantine army had largely lost the initiative; it could not repel the Arab raids by facing them in a pitched battle near the borders; consequently, a guerrilla defensive strategy was used instead. The Byzantines tried to exploit their limited resources and geography to their maximum advantage. As a result, the once-linear system of defence changed to deep zone defence. The Arabs were to be stopped, if possible, at the narrow passes of the Taurus or Anti-Taurus mountains or at the defiles of the Caucasus, but the Byzantines were hardly ever able to repel them as they entered. Therefore, the Byzantines avoided contact; instead, they followed the enemy closely, traced their routes and attempted to be constantly aware of their manoeuvres, so as to be able to avoid or ambush them at the right time, usually when they returned from the raid. The armies of the *themata* would sometimes burn the available fodder found in their province if they knew that the enemy would overrun it, aiming to put extra pressure on the enemy cavalry. As the Arabs entered Byzantine territory they found numerous fortresses with garrisons along the main strategic routes. These could provide a hindrance, especially in mountain passes, and could either significantly delay the enemy or become a potential threat if they were by-passed. In the ninth century, some narrow passes became independent administratively. These were known as *kleisourai*, and they facilitated a quicker response either to block a passage or to warn about impending raids. As a final counter-measure, the Byzantines would launch counter-raids, aiming at destroying the Arabs' resources and potential, at retaliation or at providing a distraction.[8]

Scholars have recognized three different zones of defence spread across Asia Minor: the first was the outer zone, which extended across the Taurus Mountains, and it is often interpreted as a devastated no-man's land that was constantly subjected to destruction and hostilities. Behind the outer zone lay the middle zone, which provided a second line of defence in the centre of Asia Minor and was full of fortified towns and fortresses. The final zone was in the west and northwest area where enemy raids seldom penetrated. Consequently, this was used mainly for agriculture, trade and financial growth in order to compensate for the extensive territorial losses of Egypt and Syria.[9]

The fact that warfare became endemic and standardized, in connection with the increase of literary production from the ninth century onwards, facilitated the appearance of a number of military manuals such as the *PS, LT* and *DV*. The latter is solely dedicated to frontier warfare and is the most comprehensive source on Arabic and Byzantine guerrilla tactics employed.[10] Nevertheless, all three share a common characteristic: they all present a Byzantine guerrilla strategy and mentality as well as an army suitable to undertake such a role, that is, an army which was more or less unspecialized and did not undertake specific tactical roles.

The problem with this strategy was that although it was effective in dealing with enemy raids, it entailed a considerable investment in terms of manpower and resources. The fact that many sites were raided and devastated before the enemy was confronted, and sometimes without success, seems to be a major reason why Byzantium remained relatively poor. The results of this strategy seem to suggest that life in the Arab–Byzantine frontier was very grim. However, recent archaeological evidence seems to point towards a more dynamic and complex situation.

Asa Eger has argued that although raids and warfare were dominant in the frontier, this area was by no means a no-man's land. There is evidence that the roads were maintained; that there was growth, continuity and a local economy; that trade connections and activity existed among Arabs and Byzantines; and finally that a number of fortresses and towns either on the plain or in the mountains were repaired, re-inhabited or built from scratch.[11]

While the basic features of Byzantine warfare remained unchanged from the eighth to the early tenth century, a number of significant developments occurred first during the reign of Leo VI (886–912), but more manifestly from 920 onwards. More specifically, among the numerous raiding campaigns, which had little long-term effect, a number of strategic annexations of territory started to take place, some of which allowed certain *kleisourai* to be upgraded to *themata*. In addition, the Byzantines abandoned their passive stance against the Arabs, and as soon as they had secured their Western front, they took the initiative in the struggle and assumed a more offensive role. Campaigns into enemy territory became significantly more frequent, and the Byzantine army penetrated deeper than it had ever done in the previous period of three centuries. Finally, the Byzantine army started to change to meet these new challenges. New specialized units appeared and new formations were invented. Consequently, these years mark the first steps in the abandonment of a strategy of raiding and defence, in favour of aggression and annexation. It is exactly in this military context that the *ST* was compiled, and its contents present an evolved army ready to employ the traditional guerrilla strategies, as well as support and undertake regular offensive campaigns, sieges and pitched battles.[12] It is worth highlighting these developments as we go through the main events from 900 to 950, putting them into context and finally drawing some conclusions.

First developments: the reign of Leo VI (886–912)

The reign of Leo VI is usually described as one which included numerous defeats, and Leo himself as somebody who had no clear plan for the eastern front or who was indifferent to and ignorant of military threats. However, some scholars have reviewed his reign in a more positive light, arguing that Leo VI did his best to respond to the situation of his time and that his defeats, although extensive, had little or no long-term effect.[13]

The Byzantines were largely unable to take the offensive at this time. The khan of the Bulgarians, Symeon I (893–927), was the biggest threat and a constant trouble for Byzantium from 894 to 902. In one of his campaigns in 896, Symeon reached the walls of Constantinople itself. Leo VI was in such a desperate situation that he was compelled to arm Arab prisoners of war to assist in the defence of the city. The hostilities ended with an ignoble treaty for Byzantium, and despite what was agreed, the Byzantines had to protest to the Bulgarians about the occupation of thirty Byzantine fortresses.[14]

In the east, the Arabs pressed the Byzantines hard. In addition to the usual raids by land, the Arab fleet now became a menace in the sea. From 898 to 904 the Arabs won several naval battles and looted a number of coastal towns and cities such as Lemnos (902–903) and Thessaloniki (904). The Arabs had no interest in permanently occupying these sites; they merely pillaged and destroyed them for the collection of booty and prisoners. Until 910 the Byzantines were unable to deal with this threat and to prepare a counter-attack with long-term results. They only confined themselves to launching a number of counter-raids as retaliation, attacking Cyprus, north Syria and other places.[15]

The first serious Byzantine campaign took place in 911. This expedition was different in that it was not launched aiming to raid Arab territory, but with a view to re-conquer Crete. Had the expedition proved successful, the Byzantines would have annexed lost territory and would have taken a major step in pacifying the Aegean, since Crete had turned into a significant pirate base.[16] Nevertheless, the long-term effects which military means did not manage to produce were achieved through diplomacy. Leo VI succeeded in winning over Manuel, an eminent Armenian chieftain, and thus managed to annex Kamacha and Keltzini, the addition of which allowed the surrounding area to be raised into the *thema* of Mesopotamia.[17] This small annexation was the first step in a series of developments which had a long-term effect for the Byzantines and which led to further expansion and interference in the east.

First developments: the years 912–927

In 912 Leo VI died, and the throne passed to his brother, Alexander, who only ruled for a year. After Alexander's death, a regency was formed to take care of matters, since Constantine VII was only a child at that time. Byzantium was still in a difficult situation, compelled to face the Bulgarian and Arab threat simultaneously.

From 913 to 920 Symeon continuously attacked Byzantium. His strategy involved raiding the cities of Macedonia and Thrace, most notably Adrianople and the environs of Constantinople. Since he was unable to take the capital itself, Symeon confined himself to receiving payments and gifts from the Byzantines in exchange for his withdrawal. In response to this threat the Byzantines signed a peace treaty with the Arabs. lasting from 917 to 922, so as to focus their manpower and resources in the west. This did not produce the expected results, however; the Byzantines were twice defeated by Symeon in the battles of Achelous and Katasyrtae (917), and the Bulgarians continued to menace the Byzantines until 924, well into the reign of Romanos I (920–944) who had in the meantime taken the throne for himself.[18]

The eastern front was a constant problem for the Byzantines. Raids were organized and launched mostly from the very important cities of Tarsus and Melitene. Such raids were either seasonal or specifically conducted in response to Byzantine activities. One of these raids was undertaken by al-Ḥusayn Ḥamdān (913). It was the first raid of the Hamdānids into Byzantine territory and a harbinger of the victories that Sayf al-Dawla was to enjoy over the Byzantines in the next decades. Although at this time Ḥusayn was merely the governor of the border garrisons of Mesopotamia, he had already shown his desire to acquire more authority in the whole district by interfering with the allocation of tax revenues, which went beyond his authority.[19]

Although the Byzantines suffered at the hands of both Arabs and Bulgarians, this period saw a number of developments which had some positive long-term effects for Byzantium.[20] The first was the Byzantine interference in Armenian affairs. The control of Armenia was crucial to both major powers in the area. The Byzantines preferred to hold the Armenians to their sphere of influence because they could use them as a buffer against the Arabs, while at the same time control the major routes that passed through their territory. Hence, the Arabs would be unable to cause a serious threat in the southern part of the frontier, and the Armenians could prove an indispensable ally for a Byzantine offensive. The Arabs did not want another hostile force on their borders, and the control of passes was an important element in Arab–Byzantine warfare. From 885 the Arabs had managed to secure their influence in Armenia by crowning Prince Ashot king of kings. However, his heir was more sympathetic to the Byzantines, and from 908 to 909 he was at war with the Arabs. Taking advantage of the revolt of al-Ḥusayn Ḥamdān, who turned his back on Baghdad and revolted against the caliph al-Muqtadir, the Byzantines from 915 to 919 supported Ashot II and the Armenians gained independence from the Arabs. Second, not only did the Byzantines secure their influence in Armenia until 928, but also they seem to have annexed some territory by 916, enough to allow the elevation of Lykandos from a *kleisoura* to a *thema*.[21]

The third important development was a series of events which facilitated the gradual turn of the tide in favour of the Byzantines in the eastern front. In 922 John Radinos managed to deliver a significant blow to Arab piracy in the Aegean by defeating Leo of Tripoli, who was responsible for the sack of the Thessaloniki back in 904. In addition, from 923 onwards, the Abbasid Caliphate entered into a long period of gradual decline. On one level, the nomadic tribe of the Quarmatians of Bahrain started to attack the Caliphate, conducting incursions which could not be effectively challenged for at least five years. On another level, the Abbasids were already weakened by civil wars and by internal affairs: Bagdad was in a terrible financial situation and had every difficulty in paying and sustaining its army.[22]

Nevertheless, the major development that enabled the Byzantines to achieve their counter-attack in the east was the neutralization of the Bulgarian threat in the west. In 924 Symeon prepared another expedition against Byzantium and negotiated an alliance with the Fatimids, who were to assist in the siege of Constantinople with their fleet. Romanos I became aware of these plans and managed to win over the Fatimid caliph al-Mahdi, signing a peace treaty with him.[23] In spite of this setback Symeon invaded anyway and, pillaging as he went, encamped in front of Constantinople. Unable to conquer the city, Symeon asked for a peace treaty and when one was concluded, he retired to Bulgaria. Symeon did not bother the Byzantines with another incursion, and in 927 he died. After the death of Symeon, Bulgaria was exhausted from fighting since its neighbours found the perfect opportunity to invade. His successor, Peter, had no other option but to ask for a long-term peace from the Byzantines, which was given after he agreed to marry Romanos' granddaughter, Maria, in October 927.[24]

The watershed of 927 and the years up to 944

The death of Symeon and the marriage of Peter and Maria mark a significant turning point which greatly facilitated the evolution of Byzantine warfare. Byzantium had more or less secured its western front, the Bulgarians were not a threat anymore and Constantinople was able to focus its entire manpower on fighting the Arabs and on dominating the eastern front with long-term results.[25] Byzantium now steadily undertook the offensive in the east and chose to expand its influence and territory by its own initiative. The Byzantines knew of the Caliphate's troubles and weaknesses; therefore, a new spirit of confidence arose after the conclusion of the Bulgarian wars.[26] An oration written in 927 records this optimistic spirit: 'So all things made new and sparkling, and hymn and glorify the cause of this. Only the sons of Hagar mourn and shall mourn, who are bereft of heart at the mere echo of our concord'. In addition, Ibn al-Athīr records that, in 925, Romanos demanded tribute from the Arab cities of the frontier, stating that 'I know exactly the weakness of your governors'.[27] The Byzantine actions after 925 seem to confirm this mentality.

In the light of this spirit, the Byzantines seemed to have had a general plan in mind after the Bulgarian wars.[28] That is not necessarily to say that there was a grand strategy or a higher motive behind every move but that their goals were more or less clearly dictated by a sense of realpolitik, common sense, knowledge of their capabilities and a long-acquaintance with the enemy and his strategy. Furthermore, almost all the initiatives came from Constantinople itself and were executed by the *domestic* of the *scholae*, John Kourkouas. There was continuity too, since, in many respects, Constantine VII continued the policy of Romanos I Lekapenos. Finally, it is obvious that after 925 there were repeated attempts to annex or to extend the influence of the empire to certain key cities which were the very bases from which Arab raids were launched, like Melitene and Theodosioupolis, or to strategic fortresses like Marash, Samosata and al-Hadath, all of which aimed at neutralizing the enemy threat and at acquiring a foothold in strategic areas.[29]

These new circumstances, and the more focused attempts that Byzantium undertook to achieve its primary goals on the eastern frontier, whether these led to annexations, treaties, sieges or defeating Arab armies, meant that the Byzantine offensive operations were multiplied. A new army was required to fulfil these roles, an army which would be able to undertake specialized roles, an army in which the *tagmata* would play a major role. A strong, disciplined, specialized infantry and cavalry were essential for these tasks, and the need to support and protect each other during the march, battles and sieges was crucial. Gradually, it became more and more essential to stand and fight in a disciplined formation, something which was not the major priority of the armies of the *themata*. Consequently, scholars have argued that it was exactly at this time that the Byzantine army evolved and took the form of the army presented in tenth-century military manuals, able to fight in the hollow-square formation, where the infantry served as a mobile operation base for the cavalry, able to field specialized troops such as heavy, light and medium infantry; *menavlatoi*; *kataphraktoi*; lancers and horse archers.[30] In contrast to earlier military manuals, such as *LT*, the *ST* is the first treatise to record such evolutions and specialization of formations and troops.[31]

However, the Byzantine offensive did not achieve its goals overnight. Although the Bulgarians were no longer a threat, a number of new challenges appeared that troubled the Byzantines. In 941, the Byzantines were faced with the invasion of the Rus, which compelled them to abandon the east for some time and to march back to Bithynia to repel them.[32] But the biggest challenge for the Byzantines were the Arabs. Apart from the usual seasonal raids, now mainly coming from Tarsus (926–931), the Byzantines had to face incursions which were organized, led and conducted by new powers. In the period 927–932 Mu'nis, the leader of the Abbasid army, was responsible for extra incursions against Byzantium.[33] In addition, since the caliphate was in a weak financial and military situation, its caliph, al-Muqtadir, responded to the Byzantine thrust in the east by making Mesopotamia an autonomous district, similar to Tarsus. In 930–931, the caliph decided to grant an almost autonomous north Mesopotamia to the Ḥamdānids.[34] This year would mark the active involvement of the Ḥamdānids in frontier warfare, which was to disrupt Byzantine operations until the 960s. Three major figures of the family – Saïd ibn Ḥamdān, leader of the Ḥamdānids and governor of Mesopotamia from 931–934; Nasir al-Dawla, leader of the family and governor of Mesopotamia from 934; and his brother, Sayf al-Dawla, who for now acted as a deputy of Nasir – inflicted significant blows to the Byzantines. In the period 931–940 they conducted raids into Byzantine territory; they repelled Byzantine expeditions against Melitene, Samosata, Arsamosata and Armenia; and they brought part of the latter to their control, occupying key fortresses and towns which supervised the main routes from Armenia to Mesopotamia and from Arsan to Taron.[35]

Luckily for the Byzantines, however, both the Abbasids and the Ḥamdānids were weakened by internal strife and crises. The Byzantines took advantage of the fact that the Arab world was divided which, in combination with the elimination of the Bulgarian threat in the west, allowed the Byzantines to take the initiative and to launch a counter-attack against the major Arab frontier cities.

The first main goal of the Byzantines was Melitene, and repeated attempts were made to capture it. It was a major frontier city and one of the three bases from which raids were organized and launched. It was under the authority of an independent emir who recognized the overlordship of Baghdad. However, Melitene was not only dangerous because the main policy of its emirs was to raid Byzantium, it was also strategically placed close to the Anti-Taurus Mountains, allowing the Arabs to control the plains and neutralize the natural defences of Anatolia.[36] The first Byzantine attempt against Melitene was undertaken in 926. John Kourkouas and Melias managed to break into the city for a short time, but they were repulsed and then confined themselves to pillaging its environs for sixteen days. The city turned to the caliph, al-Muqtadir, to request assistance, but none came, since the latter was occupied with internal affairs and with the extensive raids of the Quarmatians. Melitene came to terms with the Byzantine army, and a non-belligerent agreement was concluded. The emir was to keep his position but promised to maintain peace and not to raid against Byzantium.[37]

The fact that Byzantium saw annexation as a last resort and that it preferred to demilitarize enemy cities meant that the latter would come to terms with the empire easily, but this quick end of hostilities did not always guarantee long-term results. Melitene changed sides two times (916, 930) and was unsuccessfully besieged another two (928, 929) before it was finally captured in 934. At that time the Abbasids were occupied with a revolt which succeeded in overthrowing the caliph, al-Qāhir (932–934). In this context Kourkouas marched to Melitene for the last time, leading a considerable army of *themata*, *tagmata* and Armenians. After the city surrendered, it was annexed and turned into a *kouratoria*. The fall and annexation of such a major city came as a surprise to the Muslim world, and it was more than clear now that the Byzantines had come to stay.[38]

The imperial army made attempts against the other major frontier city, Theodosioupolis, which was also a base for the organization of raids. In 931 the city was taken by John Kourkouas, but once more it was not annexed; instead, a tribute and a non-belligerence treaty was imposed on it. As was the case with Melitene, the city changed sides again, and in 939 the Byzantines launched an unsuccessful expedition to conquer it. Theodosioupolis, however, would only be annexed in the reign of Constantine VII.[39]

Although the Byzantines did not secure Theodosioupolis, they managed to expand towards Armenia and Mesopotamia by capturing a number of key fortresses and towns. Around 937–938 the Byzantines acquired some territory from the emirate of Tarsus, which was extensive enough to allow the elevation of Seleucia from a *kleisoura* to a *thema*, while in 939 they annexed Arsamosata.[40] As was characteristic of Byzantine expansion at this stage though, most defeated cities and fortresses did not follow the fate of Arsamosata. Instead, they were either given to Byzantium's allies or they were sacked, forced to demilitarize and to sign a treaty of non-belligerence. The Byzantines temporarily occupied Samosata and/or Chimchat (927, 931, 936), Khelat and Bitlis (928), Perki (931), Marash and Halad (938), while the town of Mastatum and the area north of Araxes were given to the Iberians (930–931).[41] Their success was partly owed to the antagonism between the Ḥamdānids, the Fatimids and the Abbasids, as well as to the incursions of the Quarmatians and of the Qushayr and Numayr tribes.

The final years of Romanos' I reign saw the Arabs even less able to resist the Byzantine advance. The Abbasid caliph al-Muttaqī had to face the mutiny of Turkish troops (941) which now threatened his throne and compelled him to abandon Baghdad. The Ḥamdānids were called to help, and in 942 Nasir al-Dawla managed to enter Baghdad. After their successful intervention in Baghdad, the Ḥamdānids re-established themselves in their territory. Nasir was named governor of Mesopotamia and north Syria, while Sayf managed to occupy Aleppo and became its emir (944).[42]

The occupation of the Abbasids and the Ḥamdānids in Iraq, North Syria and Mesopotamia gave a golden opportunity to the Byzantines to push forward and to

raid from 942 to 944 with no serious opposition, save for Sayf al-Dawla's intervention when he defeated the Byzantines and raided the *thema* of Lykandos (944). The
Byzantines now raided and attacked the heart of Mesopotamia, directly challenging Ḥamdānid rule: Hamus, Arzan, Dara, Martyropolis, Ra's Ayn, Nisibis, Marash
and Bragas were all captured and looted. The Byzantines did not annex these cities;
rather, they most probably imposed some kind of peace treaty before abandoning
them. Last but not least, John Kourkouas marched to Edessa which came to terms
with him; its emir handed over to the Byzantines a famous relic, the *mandylion*,
and promised to stop hostilities (944). The Byzantine army had not appeared so
far east for three centuries.[43]

Constantine VII and the years 945–950

Constantine VII came to the throne in a very favourable context. Not only did the
Byzantines have the upper hand and the initiative in the eastern front, but also the
Ḥamdānids had no interest in frontier warfare from 945–948. Sayf al-Dawla was
chiefly occupied in achieving consolidation in Syria and, most importantly, taking
Damascus from the Ikhshīds of Egypt. After his unsuccessful attempts to conquer
the city, Sayf turned all his attention in subjugating Barzūyah, a fortress between
Tripoli and Antioch, which was ruled by a Kurdish governor, Abū Taghlib
al-Kurdī. At the same time, Nasir al-Dawla was caught in the struggle to invade
Iraq. The occupation of the Ḥamdānids elsewhere gave the chance to Constantine
VII to consolidate his authority and to follow the policy of his predecessor. Since
Arab resistance was very limited, the Byzantines attacked Marash and al-Hadath,
which were conquered and demilitarized and a non-belligerence treaty was
imposed on them.[44]

The years 949–950 mark a terminus for the Byzantine–Arab wars. The unsuccessful campaign against Crete in 949 brought a change in Byzantine frontier
policy. From this year onwards, Constantine VII focused his entire resources in
the east and tried to compensate for his failed expedition. For a decade the
Byzantines pushed farther and farther into the frontier every year and were now
more eager to raid the heart of the Ḥamdānid emirate. Key cities and fortresses
like Marash, al-Hadath, Martyropolis, Diyār Bakr and Aleppo were stormed and
sacked, while others which had been taken several times in the past were now
annexed, such as Theodosioupolis in 949 and Samosata in 958.[45] This Byzantine
thrust after 950 also brought administrative changes: the *strategos* started to
lose his supreme authority as the *themata* were no longer important in counter-
ing the raids, since these gradually started to fade or to become weaker. The
tagmata were now moved towards the frontier, and the *themata* were subordinate to the *katepanates* under a *doux* or a *katepano*.[46] All these set the stage for
the final push of Nikephoros II and John Tzimiskes which was to follow in the
960s and 970s.

But 949 did not mark a change only for Byzantium. Sayf al-Dawla, who was
convinced after his recent failures that it would not be an easy thing to extend his
influence in Syria at the expense of the other Arab governors, turned his attention

to the frontier and engaged in large-scale punitive raids in the borders and heart of the empire. Furthermore, he was compelled to take the defence of the frontier more seriously, as the Byzantines pushed farther and farther into his territory. As a result, he managed to become infamous among the Byzantines, inflicting some devastating defeats on them. Sayf's objective was to reverse the situation whereby the capture of Melitene and the surrounding area had made the Byzantines masters of the passes. The demilitarization of key fortresses, like al-Hadath, made matters worse, as it exposed vulnerable sections in the south and gave the Byzantines more initiative. Sayf would try to focus on controlling al-Hadath and do his best to launch regular retaliatory raids in Byzantine territory, as well as to compel the governors to revoke their treaties with the Byzantines.[47]

Reflections and conclusion

By taking the events discussed here into consideration, it is important to reflect first on what enabled this steady Byzantine expansion after the 920s and what the character of this expansion was. With regard to the first matter, it cannot be overlooked that the neutralization of the Bulgarian threat allowed the Byzantines to concentrate their manpower and resources in the eastern frontier. The fact that the Byzantine army started to evolve and change must have also played a part in favour of the Byzantines. However, the decline of the Abbasid Caliphate played an even greater role. It is vital to note that almost all major Byzantine successes coincided with periods of instability, revolt and internal strife.

By 925, the Abbasids had entered a crisis from which it was impossible to extricate themselves. The caliphate had immense trouble finding revenues for its troops, while at the same time new political and demographic elements appeared on the scene.[48] The Quarmatians formed the largest nomadic migration since the seventh century, and at first, the caliphate had no other option but to deal with their extensive raids and pillaging. These nomads played a major role in the historical scene of Syria and Mesopotamia during this period and up to the eleventh century.[49] Apart from the Quarmatians, the Abbasids had to deal with certain alien groups which at times served the caliphate and at times revolted against it: the Daylamite and Turkish soldiers, as well as the Buyids, put an immense pressure on central authority and kept its army occupied at the expense of the war against the Byzantines. In addition, the semi-independent emirs and governors of certain emirates and cities were always a potential threat. They were usually involved in wars between one another for the expansion of their influence, and others like the Ḥamdānids were actively involved with the affairs of Baghdad while trying to pursue their own interests. Allowing a district to be independent had some financial and administrative benefits for the Abbasids: the independent district would be responsible for the collection of taxes and the upkeep of garrisons, defence of the frontier and engaging in raids. But this also meant that its governors could turn their back on central authority or enter into individual negotiations with the Byzantines.[50]

The Ḥamdānids, who were responsible for many Byzantine defeats and the defence of Mesopotamia and north Syria, were of course not in such a precarious situation, but their estates and power were limited. They, too, had to face the hostilities of Bedouin tribes and were sometimes forced to pay tribute. Furthermore, Nasir was not too actively involved in the frontier and was always occupied in pursuing his interests in Iraq and Bagdad. The role of frontier warfare fell to Sayf, who was also reluctant to play the role and was only seriously involved with it after all his attempts on Damascus and Syria had failed.[51]

With regard to the character of this expansion, although we are indeed talking about a Byzantine offensive, it is important to underline that its motive was essentially defensive. The Byzantines sought to occupy territory only if it was absolutely necessary or when there was no other option.[52] While their main concern was to improve their strategic position by extending their influence on the control of passes and by halting the enemy raids, this was seldom done with annexations. Even one of the main targets, Melitene, repeatedly entered into peace treaties and pledges of neutrality before it was finally annexed.

It is obvious from this discussion that the Byzantines preferred not to annex territory that was difficult to control or was too vulnerable; instead, they preferred to acquire only key fortresses and towns, while the rest were either given to allies, like the Iberians, or were demilitarized, forced to pay tribute and cease hostilities.[53] The *DAI* records how annexations primarily served a defensive purpose; we read that 'if these three cities, Khelat, Arzan and Perki, are in the possession of the emperor, a Persian army cannot come out against Romania, because they are between Romania and Armenia, and serve as a barrier and military halts for armies'.[54] Therefore, the primary character of Byzantine expansion was to extend the influence of the empire and gain access to key areas and cities, first by demilitarizing them and making them tributary, and if this was not possible, through annexations. This gives a more demilitarized character to the Byzantine frontier than we usually think of it.[55]

In this light, most of the campaigns which did not result in annexations played the roles noted earlier, while at the same time helped to show the flag deep into Arabic territory, which could be of great importance in achieving diplomatic results by making the weakness of Baghdad obvious and by convincing governors and emirs to surrender and enter into non-belligerence agreements.[56]

Although Byzantine expansion was much more marked during the reigns of Nikephoros II Phokas and John Tzimiskes due to major annexations, the roots of their achievements are to be found in this period.[57] The gradual developments described earlier mark the military context which inspired the compilation of the *ST*. The latter presents an army and strategy able to operate in such a context, namely, to fulfil both guerrilla and tactical offensive warfare. Consequently, it comes as no surprise that the innovations of the *ST* were adapted and further evolved in military manuals like the *PM* and the *TNO*, which were both authored by eminent generals who were active in the later years of Byzantine expansion, namely in the second half of the tenth and the beginning of the eleventh century.

Notes

1　Asa Eger 2014: 4–5; Cheynet 2001: 57; Haldon and Kennedy 1980: 79–83.
2　See for example: Stouraitis 2009: 47–54; Lilie 1976: 287–338; Haldon 1990b: 208–53.
3　The literature for the *stratiotai*, the *themata* and military service is vast; some comprehensive works include: Karayannopulos 1959: 37–97; Ahrweiler 1960: 1–24; Oikonomides 1972: 340–63; 1988: 62–86; Toynbee 1973: 224–52; Haldon 1979; 1993: 1–67; Whittow 1996: 170–4.
4　Haldon 1984; Kühn 1991; Kolias 1994.
5　Cheynet 2001: 60; Haldon and Kennedy 1980: 79–86, 101–6.
6　Trans. Asa Eger 2014: 280–1.
7　Kennedy 2001: 105–7; Haldon 1999: 39–41, 755–6; Haldon and Kennedy 1980: 113–16; Bonner 1996: 69–107; Asa Eger 2014: 5; Canard 1951: 716–17; Bikhazi 1981: 429.
8　Stouraitis 2009: 54–169; Lilie 1976: 339–60; Tougher 1997: 77–9; Cheynet 2001: 59; Whittow 1996: 175–81; Asa Eger 2014: 255; Holmes 2002: 87; Decker 2013: 137–43. See also Haldon 2013: 380–6 for a study of fortresses, information, guerrilla strategy and physical context in the Middle Byzantine Period with some archaeological evidence.
9　Lilie 1976: 339–60; Haldon 1999: 62. For the Taurus and Anti-Taurus Mountains see: *TIB:* ii, iv, v.
10　Dagron and Mihăescu 1986: 139–287; Theotokis 2012: 5–15.
11　Asa Eger 2014: 6, 250, 262; 2011: 1–13. For these aspects of the frontier see also Ahrweiler 1971: 209–30; Obolensky 1971: 303–13; Haldon and Kennedy 1980: 87–100.
12　See Chapter 5 for more information.
13　See for example: Vasiliev 1935–1968: ii.i.115–25; Canard 1951: 722–3; Tougher 1997: 163–6, 193. On Leo VI see also: Riedel 2018.
14　TC, 359–62; Tougher 1997: 173–83; Vasiliev 1935–1968: ii.i.126–32. For Bulgarian–Byzantine relations in the course of the tenth century see: Kyriakis 1993: 133–59, 211–16.
15　TC, 366–9; Vasiliev 1935–1968: ii.i.132–77; Tougher 1997: 183–93; Eickhoff 1966: 256–61. For a study of Byzantine naval warfare and the fleet in this period see: Ahrweiler 1966: 45–135 and for a study of Byzantine islands: Malamut 1988. For Thessaloniki see: *TIB:* i, vi, xi, xii and for Lemnos: *TIB:* x; Malamut 1988: i.231.
16　Tougher 1997: 191–3; Vasiliev 1935–1968: ii.i.196–216; Canard 1951: 723.
17　*DAI*, 50.117–32; Runciman 1929: 121; Tougher 1997: 40. For Kamacha see: *TIB:* ii.73, 99, 120, 233, 281 and for Keltzini, ii. 101, 105, 120, 253.
18　TC, 385–91, 400–5; Symeon Magister, 135.23; c.f. Skylitzes 204–5 with Grigoriou-Ioannidou 1983: 123–48. See also Runciman 1929: 50–7, 84–90. For Adrianople see: *TIB:* vi, xii, for Achelous, iii, and for Katasyrtae, xii.85, 116, 126, 218, 443f., 675.
19　Runciman 1929: 124–5; Vasiliev 1935–1968: ii.i.229–30, 238; Bikhazi 1981: 299–300. For Tarsus see: *TIB:* v. and for Melitene, vii.89, 92, 94–6, 99, 102, 105.
20　Haldon 1999: 40–1.
21　*DAI*, 50.135–66; Constantine VII, *De Thematibus*, 12; Vasiliev 1935–1968: ii.i.225, 232–4; Runciman 1929: 126–33; Oikonomides 1972: 350; Toynbee 1973: 146, 257–8; Pertusi 1952: 144–6; Bikhazi 1981: 308–9. For Lykandos see: *TIB:* ii.41, 45–6, 82–5, 88, 93, 97f., 108f., 129, 131, 222f.
22　TC, 405; Runciman 1929: 135–6; Vasiliev 1935–1968: ii.i.249; Bikhazi 1981: 359. For the Quarmatians see Cappel 1994: 114–18; von Sievers 1979: 230–3; Kennedy 2004: 285–90.
23　Runciman 1929: 136; Vasiliev 1935–1968: ii.i.251–5; Canard 1951: 726.
24　TC, 405–15; Runciman 1929: 90–8.
25　For studies on the Balkan frontier see: Madgearu 2013; Stephenson 2000: 18–53.

26 Canard 1951: 718–19; Shepard 2001: 19–20; Vasiliev 1935–1968: ii.i.261–2; Runciman 1929: 135.
27 Dujčev 1978: 281; Vasiliev 1935–1968: ii.ii.148; Runciman 1929: 137. For the attitudes of the Byzantines towards their enemies through orations and epistolography see: Kolia-Dermitzaki 1997: 213–38.
28 Haldon 1999: 42–6; Canard 1951: 716–22.
29 Shepard 2001: 20, 34–40; 2012: 505–37. For Theodosioupolis see: *TIB:* ii.85–8, 103, 114, 116, 128, 164, for Marash, v.342, for Samosata, ii.42, 49f., 79, 81, 85, 87, 89 A. 283, 90, 125f. 204 and for al-Hadath, v.49, 54, 57.
30 TC, 459–60; Shepard 2001: 37–8; McGeer 1988: 135–45; 1992: 228; 1995a: 262, 272–5, 283–5; Haldon 1999: 217–20; Whittow 1996: 323–4; Decker 2013: 152–6.
31 See Chapter 5 for more information
32 TC, 423–6; Runciman 1929: 111–13. For Rus and Byzantium in the tenth century see: Franklin and Shepard 1996: 91–168. For Bithynia see *TIB:* ix, xii.
33 Runciman 1929: 138–41; Vasiliev 1935–1968: ii.i.258–60, 263–8; Bikhazi 1981: 414–15; Canard 1951: 734.
34 Bikhazi 1981: 414.
35 Vasiliev 1935–1968: ii.i. 277, 282–90; Whittow 1996: 318; Runciman 1929: 143–4; Bikhazi 1981: 429, 578–85; Canard 1951; 736–7, 741–7. For Arsamosata see *TIB:* ii. 78f.
36 Cheynet 2001: 59–60; Whittow 1996: 310; Haldon 1999: 725–32; Runciman 1929: 142; Haldon and Kennedy 1980: 109–12. The other two bases were Theodosioupolis and Tarsus, see pp. 19–20. For the importance of Tarsus see: Bosworth 1992: 270–86.
37 TC, 415–16; Runciman 1929: 137; Vasiliev 1935–1968: ii.i.258–9; Bikhazi 1981: 366–7; Canard 1951: 732.
38 TC, 416–17; Runciman 1929: 137, 139, 141–2; Vasiliev 1935–1968: ii.i.264, 267–70; Canard 1951: 733–6; Bikhazi 1981: 413–16, 428–9; Whittow 1996: 318.
39 Vasiliev 1935–1968: ii.i.284–5; Bikhazi 1981: 583–5; Canard 1951: 744–5; Whittow 1996: 310.
40 Constantine VII, *De Thematibus*, 13; Runciman 1929: 146–7; Oikonomides 1972: 150; Toynbee 1973: 258; Pertusi 1952: 147–8; Asa Eger 2011: 140; Whittow 1996: 318; Canard 1951; 745.
41 *DAI*, 45.50–175; Asa Eger 2011: 114, 147; Canard 1951: 733–7; Vasiliev 1935–1968: ii.i.260–1, 264, 266, 277; Bikhazi 1981: 580–1; Runciman 1929: 138–41. For Bitlis see: *TIB:* ii.52, 55.
42 Vasiliev 1935–1968: ii.i.290–1, 304–6; Bikhazi 1981: 474–505, 522–5, 607–10; Kennedy 2004: 273.
43 TC, 426–8, 432; Runciman 1929: 144–7; Vasiliev 1935–1968: ii.i.295–6, 304–6; Bikhazi 1981: 586, 609–10; Whittow 1996: 321; Canard 1951: 748–53; Kennedy 2004: 273. For Hamus see: *TIB:* v.56, 72, 82f., 135, 269, for Arzan see Theodosioupolis, for Dara, ii.87, for Martyropolis, ii. 87, 128, for Ra´s Ayn, xiv, for Nisibis, ii.87, 92, 235, for Edessa, ii.66, 87, 96, 106, 110, 235. For the *mandylion* see Guscin 2009.
44 Vasiliev 1935–1968: ii.i.316–18; Bikhazi 1981: 548–58, 611–23, 657–9; Canard 1951: 761.
45 Shepard 2001: 34; Whittow 1996: 322; Bikhazi 1981: 787, 843–4, 854–67; Canard 1951: 762. For Diyār Bakr see: *TIB:* ix.123, 191 and for Aleppo, v. 56ff., 60ff.
46 Ahrweiler 1960: 46–63; Cheynet 2001: 59–69; Holmes 2002: 88; Oikonomides 1971: 285–302; Kühn 1991: 60–6, 158–69; Krsmanović 2008; Andriollo 2017: 76–318.
47 Bikhazi 1981: 609–11, 732–5; Haldon and Kennedy 1980: 108–9; Kennedy 2004: 273.
48 Whittow 1996: 329; Bikhazi 1981: 450; von Sievers 1979: 212–44; Kennedy 2004: 186–90; Canard 1951: 719.
49 Cappel 1994: 114, 117, 129; Asa Eger 2014: 265.
50 Bikhazi 1981: 450; Whittow 1996: 329–32.
51 Shepard 2001: 35; Kennedy 2004: 269–71, 274; Bikhazi 1981: 705–10, 777; Whittow 1996: 333–4.

52 Shepard 2001: 20, 22–5; Cheynet 2001: 59.
53 Shepard 2001: 26–30, 36–9.
54 *DAI*, 44.125 (trans. Jenkins, p. 205).
55 Shepard 2002: 67–82; Holmes 2002: 97–103; 2008: 149, 151–3; Cheynet 2001: 58.
56 Shepard 2001: 21.
57 For the difference in strategy between these two emperors see: Garrood 2013: 20–34. The expansion of Byzantium and its conquests from the reign of Nikephoros II Phokas onwards have been most recently studied in detail by Kaldellis 2017.

2   The sources of the *Sylloge Tacticorum*

As Vári noted in the first half of the twentieth century, 'with every Byzantine author of military manuals we must primarily inquire into the sources used, because there is hardly any Byzantine military manual which does not rely on older works, or that does not seek to extract ideas from older military treatises without compunction'.[1] The case of the *ST* is, indeed, no different, as its author took much of his material from earlier manuals, both ancient and Byzantine. The issue of imitating older sources, however, is something found not only in the *ST* but in almost every genre of Byzantine literature. Consequently, to do justice to the problem of the sources of the *ST* and to understand the concept of imitation better, one needs to put it into a broader context and to reflect on its connection not only with Byzantine literature in general but also with the education of the aristocracy.

The context of imitation in Byzantine literature

The phenomenon of imitation, or *mimesis*, in Byzantine literature is identified as an act of copying from classical and/or earlier Byzantine works, which involved using similar vocabulary or sometimes reproducing verbatim from the source, usually without mentioning the original author. Furthermore, imitation is also recognized in the adoption of the same motifs, aesthetic and style. Since plagiarism and copyright law were not issues in Byzantium, the author of a Byzantine literary work actually showed how well versed he was with the 'bibliography' of his time and therefore demonstrated and validated his value and erudition by imitating older works.[2] But imitation of older authorities was not only the norm in Byzantium, it was a concept of the medieval word in general, and it is also found in the West until the middle of the eighteenth century.[3]

The concept of imitation may seem peculiar today, but the lucky few Byzantines who were educated were exposed and encouraged to such behaviour from an early age. From the time they were young students, they worked with rhetorical textbooks, the *progymnasmata*, which instructed them to imitate older forms. A good example comes from the textbook of Theon who argues that:

> Despite what some say or have thought, to paraphrase is not without utility.
> The argument of opponents is that once something has been well said it cannot

be done a second time, but those who say this are far from hitting on what is right. Thought is not moved by any one thing in only one way so as to express the idea (. . .) but it is stirred in a number of different ways and sometimes we are making a declaration, sometimes asking a question, sometimes making an inquiry (. . .) and sometimes expressing out thought in some other way. There is nothing to prevent what is imagined from being expressed equally well in all these ways (. . .). While Homer says 'Such is the mind of men who live on earth, as the father of men and gods grants it for the day', Archilochus rephrasing the lines says 'Such Glaucus, son of Leptines, is the mind of mortal men as Zeus brings it for the day'.[4]

Consequently, from the time of their education and from literary conventions, the readers of Byzantine literature would have been crystallized in such a tradition and would have rather enjoyed recognizing the reference to ancient or older authorities.[5] Beneficiaries of such an education formed a distinctive group which was usually employed in Byzantine bureaucracy or found a place in the emperor's court or the church. In fact, according to Psellos, it was enough if somebody recounted two to three words for the educated aristocrat to recognize the reference to a classical author and admire the man for his learning.[6]

Imitation of older works and adhesion to the tradition were largely the norm due to education, but those who must also have been influential in shaping and maintaining this attitude were the patrons and readers of such literary works, and more specifically in our case of military manuals. The target group of these works and the relationship between the patron and the ghost author were paramount. The advantageous position of the patron or commissioner, who most of the times seems to have been an emperor, would have been crucial.[7] The emperors would have accepted imitation as the appropriate way to compile a manual and would have therefore encouraged the incorporation of older authors, classical or Byzantine, not only because they themselves had usually received an elite education but also because they would have been seen as protectors and continuators of the classical tradition and wisdom.

Most of the time, the same applied to those who seem to have been the readers of such works, namely the military aristocracy. This relatively small group of people was sometimes educated to a certain extent and shared the same traditional standards, which dictated that the classical tradition gave authority to any literary work. After all, these men were sometimes copyists, owners, authors or readers of literary works.[8] The military aristocracy was certainly familiar with military manuals, including ancient ones. Kekaumenos records that he had received education from the *progymnasmata*, while he and Nikephoros Ouranos were evidently versed with the works of the ancient tacticians and also advised their readers to consult them.[9] In addition, John Doukas and Alexios I are recorded as having read the works of Aelian and Apollodorus.[10]

But apart from the impact of patrons and readership, Byzantine authors had two basic things to gain from classical imitation. The first was the authority that the ancient tradition had, and the second the refined language and style of ancient texts. Authors of Byzantine military manuals were very much interested in the first aspect, since classical authority gave them the credibility which was needed to justify their content and instructions.[11]

The style and language, however, seem to have only partially interested them. This is rational if we consider that manuals were not solely works of literature, they were also supposed to be practical handbooks. Of course, every military manual was different and sometimes more or less literary. But, although some authors employed a *topos* in their preface to take pride in the fact that they had used simple and understandable language, most of the manuals have a mixed language. They incorporate both elements of Attic and/or literary phrases, but most of their material is written in contemporary military terminology or in a military slang that had Germanic, Turkish, Latin, Arabic and Iranian elements.[12] The *ST* is a characteristic example of that. On the one hand, its author used Attic Greek forms, similar to these found in Thucydides, which do not appear in his sources, for instance, ξυμμαχικὸν instead of συμμαχικὸν, ἀντιτάττονται instead of ἀντιτασσονται and παραφυλάττειν instead of παραφυλάσσειν.[13] In other passages, he uses literary phrases, such as 'with fire and sword', which is commonly found in ecclesiastical texts, already from the fourth century by John Chrysostom, as well as in ecclesiastical commentaries, *synaxaria* or lives of saints, up to the tenth century and beyond.[14] His terminology is no different – he uses common and rare Greek terms like the word μόσυνας to describe wooden towers. This word is described as archaic already in the writings of Strabo who presents its etymology; it is used by Aeneas the Tactician in the fourth century BC and again by Anna Komnene in the *Alexiad*.[15] On the other hand, next to the classical terminology, we find contemporary Byzantine terminology which is sometimes of Latin or barbaric origin like the words *klibanion, viglas, allagia, artzikidion, menavlion, kabadion*, etc.[16]

It is therefore important to take into account that a number of different factors, well-established practices and literary convention resulted in making imitation an important aspect of Byzantine literature and, by extension, of Byzantine military manuals. Despite the fact that the *ST* was actually part of a well-established literary tradition, it is usually seen as something out of the ordinary and it is criticized for its dependence on earlier sources. To remedy this view, one should look at sources of the *ST* with a fresh pair of eyes.

The sources of the *Sylloge Tacticorum*

At first sight the issue of the sources of the *ST* is uncomplicated and straightforward; its author drew extensively on earlier works, both ancient and Byzantine. The extant works which were used by the author of the *ST* have been to a great extent identified and can be divided into two categories: those found in the first half of the treatise (chapters 1–55) and those found in the second half (chapters 56–102).[17]

As far as the first half of the treatise is concerned, the author of the *ST* relied on classical works, with the *Strategikos* of Onasander holding the most prominent place.[18] Onasander (fl. 50 AD), wrote a treatise which mainly focuses on the characteristics and duties of the ideal general, as well as on just war and practical aspects of warfare. Other classical sources include Arrian's (c. 113 AD) *Ars Tactica* and *On the Tactical Arrays of the Greeks* of Aelianus Tacticus (fl. 120 AD), two military manuals which analysed the equipment and formations of Roman and Hellenistic armies, respectively.

Except for classical works, the author of the *ST* drew on earlier Byzantine works too. The most prominent of these are military manuals, including the famous *MS*, the *PS* (which latest scholarship dates to the ninth century) and the *LT*.[19] From all the manuals, *LT* seems to play the most vital role in the *ST*. The author of the *ST* had relied heavily on it, both in terms of structure and in terms of material, some of which, however, he further advanced and revised. It seems that the manual of Leo VI was used as sort of a general guide for his own work. In spite of the differences in style and form – for example, the author of the *ST* does not treat his work as a *Procheiros Nomos*, neither does he divide his work in constitutions – the *ST* preserves some similarities with the structure of *LT*. Thus, the first chapter of the *ST* is fairly similar to *LT*'s first and second constitution where Leo VI discusses the qualities of the general.[20] *LT* includes a discussion of Hellenistic armament and array, and very similar material is also found in the *ST*.[21] Furthermore, the author of the *ST* included passages that refer to the well-being of the *stratiotai*, which are copied, with slight revisions, from similar ones found in *LT*.[22] Likewise, the same seems to apply for the final part of the *ST*; chapters 76–102 are dedicated to stratagems of classical commanders, while constitution 20 of *LT* includes, among other material, stratagems and anecdotes from classical sources such as Polyaenus, Plutarch or Polybius, though not exclusively, as much of what is presented also derives from *MS*.[23]

The list of Byzantine works used by the author of the *ST* extends to other genres as well. For instance, the legal texts of the *Poinalios Stratiotikos Nomos* and *Procheiros Nomos* seem to have been employed for the drafting of chapters 50.4 and 50.6, respectively,[24] and a small reliance can also be recognized on a text which belongs to the mirror of princes, Basil I's *Hortatory Chapters*, from which the author of the *ST* used a small passage and two small phrases to supplement Onasander's material.[25] Another genre which we can identify in the sources of the *ST* is that of *anthologia* or *florilegia*, which included sayings of famous figures of both classical and Christian antiquity. These texts were very popular in Byzantium; authors of admonitory works and students of elementary rhetoric (*progymnasmata*) often mined them for material since they found plenty of moralizing sayings there.[26] The author of the *ST* depended on such works to supplement Onasander and the profile of the ideal general.[27] Therefore, although we find sayings which originally belonged to Greek philosophers and orators, such as Aristippus of Cyrene or Isocrates, these seem to have been copied from works such as John Stobaeus' *Anthologia* and from more recent *florilegia* like the late ninth-century *Corpus Parisinum*.[28] For now, one cannot exclude the possibility that the author of the *ST* might have known Isocrates directly since one of his quotes cannot be traced in existing *anthologia*. As more of the latter are edited and added to the *TLG*, however, it is very probable that this saying will also be identified as deriving from such texts.[29] Finally, the author of the *ST* may have also relied on a sixth-century AD architect, Julian of Ascalon. Julian has been recognized as a source for chapter 3, but this attribution is problematic since there is no consensus as to whether the metrological list which appears in some versions of Julian of Ascalon is original.[30] An alternative source for chapter 3 could have been a mixture of additions made by the author of the *ST* along with what is found in Heron's *Geometrica* 4 and the metrological table found in the *Peri Metron*, a small treatise interpolated into the tradition of Aelian, the dating of which is unclear, other than that it dates before the eleventh century.[31]

With regard to the second half of the treatise, most of it comprises sources which are mostly Byzantine paraphrases of earlier works. This part of the *ST* is dominated by the presence of two little-known sources: the *Apparatus Bellicus* and the *Hypothesis*. The *Apparatus Bellicus* is a work most probably dating to the ninth century, which draws mainly on the *PS* and the *Cesti* of Julian Africanus (160–240 AD), and which, among others, includes various recipes for fighting the enemy with poison, as well as for producing antidotes.[32] The *Hypothesis* is a Byzantine abbreviation of the *Stratagemata* of Polyaenus who lived in the second century AD. It dates some time before 850 and presents 356 stratagems of commanders of antiquity in a thematic order.[33] Last but not least, a small dependence can be recognized on the so-called *Cynegeticus* of Urbicius, although not undisputed,[34] and on *How to Withstand Under Siege* of Aeneas the Tactician, who seems to have been the source for chapters 70 and 76.[35]

The problem with the sources of the *Sylloge Tacticorum*

The issue of the sources becomes more complicated because the author of the *ST* does not seem to have had access to some of the earlier works directly. It has been argued that the second half of the *ST*, chapters 56–102, derive directly from a lost work which has been dated to the late ninth or early tenth century, the so-called *Corpus Perditum*.[36] Although we have to speculate to a certain extent when we talk about lost sources, the existence of the *Corpus Perditum* has been convincingly supported with parallel evidence from a comparative study of the *Apparatus Bellicus*, the *ST* and the *TNO*, with scholars only expressing certain dissent concerning the version of the *Apparatus Bellicus* that the author of the *Corpus Perditum* had used.[37]

The fact that the author of the *ST* must have drew most of his material for the second half of his treatise from a lost source raised questions as to whether the same could have applied for the first half of his work as well: chapters 1–56. The main student of the *ST*, Dain, was positive of this theory and argued that the first half of the treatise derived from another lost source which he named *Tactica Perdita*. The existence of the latter, however, was not convincingly proven with a parallel study of extant sources, as was the case with the *Corpus Perditum*. The mere proof which brought the *Tactica Perdita* to life was a number of inaccurate cross-references in the first half of the *ST*. Dain envisaged these as editorial mistakes and argued that the author of the *ST* had accidentally copied them while he was drawing material from the *Tactica Perdita*.[38]

Consequently, Dain regarded the *ST* as having been copied from two lost sources, and in a rather slavish manner. This approach is partly responsible for a certain distrust of the information of the *ST* and has led some scholars to regard the work as a forgery or as a purely theoretical treatise which did not reflect contemporary Byzantine practice. If this is the case, it probably means that the internal evidence of the *ST* cannot be taken at face value and that it is of little help to contribute to the identification of its author and dating. It is therefore worthwhile to look into the matter from a fresh perspective and to determine whether one can really trust the information of the text.

Lost sources and authorial methods: a fresh perspective

The idea that the *ST* is an extraordinary example of slavish copying and poor editorial practices becomes null when it is placed under closer scrutiny. The author of the *ST* used more or less the same working methods as other authors of military manuals, and the internal information of the text can be trusted at least to the same degree as other works of this genre, say *MS*, *LT* or the *TNO*. This will become more apparent when one moves the *Tactica Perdita* out of the equation and studies the way in which the author of the *ST* treated extant sources.

To begin with, the fact that the *ST* has seven inaccurate cross-references does not necessarily mean that these were a product of careless copying from a lost source, the *Tactica Perdita*.[39] A more careful look at these cross-references denotes that all the information that the author cites does actually exist in the manual, just not in the right order. If the author of the *ST* had indeed copied from a lost source sloppily, that would hardly have been the case.

More specifically, the first incoherence can be located in chapter 21.4, where the author states that 'the outpost which is close to the enemy will consist of (. . .) at least ten men (. . .) when it is separated from the camp at a distance of approximately three or four miles in proportion to the location of the site and the preparations of the enemy, as I have said'.[40] Despite the fact that relevant information does not appear before this passage, it does appear in the next paragraph; and we can additionally locate some passages throughout the treatise, where the author has stated that a distance above three miles is not considered safe and extra precautions are needed if it is exceeded.[41] Likewise, in the second inconsistency, located in chapter 22.5, our author reports that 'As I already said, it [is] advantageous to throw iron caltrops (. . .) all around the trench'.[42] Although this is the first time that iron caltrops are mentioned in the treatise, our author discusses them in detail later in chapter 38.12.

The next three false cross-references are slightly different since they refer to incorrect chapters. The first one can be found in chapter 39.3 where we read that 'In their saddles they should all have maces (. . .) and (. . .) two or three saddle-bags containing hardtack or flour, as we have said with more detail in chapter 20'.[43] Once again, although chapter 20 has nothing relevant to offer,

more information regarding provisions in the saddle bags is found in chapter
44.8 where the author of the *ST* explains that they should be enough for two or
three days. Similarly, in chapter 46.5 we read that 'the *tetrarchs* [will fill] the
tenth and last row, on this account they are also called file-closers. We have
discussed them in more detail in chapter 17'.[44] Despite that nothing of the sort
is found in chapter 17, the positioning and naming of the *tetrarchs* are covered
in chapter 35.9. The third similar inconsistency can be seen in chapter 49.8
where our author instructs the reader not to cross the defiles without first seizing
them with the infantry; then he adds: 'as we have said in chapter thirty-three'.[45]
Although relevant information is not provided in chapter 33, it does appear in
chapter 23. It could be that, since the two chapters have a similar numbering,
some of the copyists were careless and accidentally wrote the wrong chapter
down. However, the fact that the numbering is written in full, in the surviving
manuscript at least, and that we have also encountered a number of other inconsistencies makes it less likely so.

The final two false cross-references are slightly different from the ones noted
earlier. The first one occurs in chapter 45.26, where we read that 'If the infantry
formation amounts to 6,000 [men], the same array will follow, as has been said'.
While once again nothing similar has been said earlier, this case could very well
be a mistake of the copyist regarding the numbers. More specifically, in paragraph 22 of the same chapter our author wrote: 'If the army amounts to 10,100
[men] the same array will follow, even if the army should consist of more than
24,000 [men] and up to 10,100'. It is obvious that it is unnecessary for the number 10,100 to appear a second time here, so it does not seem wrong to replace
the second 10,100 with the number 6,000. With regard to the final case, which
is found in chapter 45.25, our author informs us that: 'To reckon this infantry
formation to a total of 10,000 < and 100 > [men], 64 [men] are also included,
namely they who are detached for the needs of the three major shield-bearing
tagmata, as stated above'.[46] Likewise, in spite of the lack of relevant information
earlier, this incoherence can be attributed to an omission, as this information
should indeed have appeared in paragraph 45.7. In any case, as it can be seen in
chapters 45–47, it is a well-established pattern of the author of the *ST* to give
this information every time he completes the treatment of each variation of the
battle formations and has nothing to do with an irrelevant piece carelessly copied
from a lost source.

Taking into consideration that all the information cited by the author of the *ST*
is actually there but in a mixed-up order, we may better explain the inconsistent
cross-references by envisaging a partial revision or re-ordering of the material,
perhaps at a later stage. Although the study of this revision can prove complicated
or with too little evidence to draw secure conclusions about it, what is of primary
importance here is that most probably no such thing as the *Tactica Perdita* ever
existed.[47]

Nevertheless, the theory of the *Tactica Perdita* must be abolished not only
because the inconsistent cross-references are irrelevant to its existence but also
because the argument which wants the author of *ST* to have used two different lost

sources with a clear division in chapters 55–56 does not hold true. Sources like the *Apparatus Bellicus*, the *PS* and the *Hypothesis* which were part of the lost *Corpus Perditum*, can also be found in the first half of the *ST*.[48] In the light of what was noted earlier and of the fact that sources in the first part of the *ST* were used indiscriminately so that relevant material in very similar order and with very similar phrasing cannot be found in other treatises before or after the *ST*, it seems rational to argue that the author of the *ST* had used only one lost source for the entire treatise (i.e. the *Corpus Perditum*) and that he mainly derived the material of the first 55 chapters from extant individual sources.

Therefore, if one was to distrust the internal information of the *ST*, this would not have been the result of a supposedly extensive copying from the *Tactica Perdita*, but of the idea that the author of the *ST* employed a slavish copying technique as a whole. Even this approach, however, does not seem to be valid. All who have studied the *ST* so far agree on the fact that the treatise was written by a single author who employed a quite free method of treating his sources. Indeed, this becomes evident when one compares how the author of *ST* and Nikephoros Ouranos have treated the same material (i.e. the treatises which constituted the *Corpus Perditum*). Nikephoros Ouranos was truer to the original text, while the author of *ST* rephrased it and added extra information or comments.[49] The same applies for the first half of the *ST*, where, at closer scrutiny, one realizes that, most of the time, the author of the *ST* applied the same method of free adaptation, with changes of sequence, omissions and additions. For example, he updated the tradition of Onasander and Julian Africanus and made it relevant to his own time and context, while the same also applies for the stratagems of Polyaenus. Last but not least, he also updated the advice of older Byzantine manuals and, based on them, he created his own original advice on equipment, tactics and strategy.[50]

This idea puts the author of the *ST* into the same category with other Byzantine authors of military manuals and makes the narrative of his work as least as credible as other treatises, like the *TNO* or *LT*. Consequently, the issue of sources and imitation holds no extraordinary place in the *ST*. The majority of the material of the treatise can be traced in extant works which makes the existence of extensive and largely unknown lost sources implausible. As most Byzantine authors of military manuals and compilation literature, it seems that the author of the *ST* had sources at his disposal in more or less the same version as we know them today, or perhaps in a slightly different or abbreviated one. Therefore, it is not really his working method that is to be blamed for the relevantly bad reputation of the treatise, but rather the preconceptions of older Byzantine scholarship and the fact that the *ST* remains vastly understudied. For example, while Dain recognized that the author of the *ST* had reproduced material from extant Byzantine manuals, he severely underestimated his dependence on these sources, which in fact cover most of the first half of the *ST*. This could have influenced Dain's judgement, who sought to invent another lost source, the *Tactica Perdita*, to account for the passages he could not trace to an extant source.[51]

In addition, older scholarship was not very developed in understanding and appreciating Byzantine imitation. Byzantine *mimesis* was considered a slavish procedure and a phenomenon which was stuck to ancient models with little concern on updating the tradition and transmitting the reality of its time.[52] Consequently, it was typical to attribute to extant Byzantine authors all inconsistencies found in their works as careless copying and to consider all adaptations and paraphrases as belonging to the pen of non-extant authors. A good example to demonstrate this can be found in chapter 55.4 of the *ST*, which is similar to the information found in *MS* 10.4.34–40 and *LT* 15.62. In the *ST* we read that 'The best time for the construction of such forts should be the months of *Panemos, Loos* and *Gorpiaios*, which the Romans call July, August and September, since the pasturage is dry and easily burned at that time', while the other manuals report that 'these undertakings are effectively employed (. . .) about July, August or September when the grass is dry and burns easily'.[53] Older scholarship would have readily assumed that the passage of the *ST* derived from a lost intermediate source. At best, it might have argued that it came from a different or an unknown version of *MS* and *LT*, but it would seldom celebrate the idea that the fact that the months appear with their classical names is a personal preference of the author of the *ST* who wanted to impress his readers with his familiarity of the classical tradition.[54] This after all, as we have seen earlier, is in accordance with his overall style and with his use of Attic Greek and classical terminology.

Thus, when one notices differences in the phrasing and style of two extant works, there are more explanations to consider than simply to accept that all of them must be attributed to a verbatim copy from lost sources. Some of these adaptations may have to do to with an innovative style or a personal preference of the author for an extant work. After all, for these changes to have existed, someone at some point needed to have deviated from the original text. But to appreciate this procedure, we need to reflect on the inter-textual relations between sources and Byzantine works and better understand the role that the past played in Byzantine society.[55] Fortunately, despite the older apologetic views on imitation in Byzantine literature, recent scholarship has not confined itself to merely identifying the sources used by Byzantine authors. Instead, it focuses on better understanding this feature of *mimesis*, providing insights about the relation of Byzantium with its tradition. For instance, Ingela Nilsson has argued in favour of a *mimesis* that was perceived as 'consciousness of historical change' or an 'invitation of beginnings', as Byzantine authors not only imitated but also criticized, paraphrased or revised their sources.[56]

Nevertheless, one should be cautious. There are times when the author of the *ST* speaks in the first person, but this is not a reference to himself, but rather a direct interpretation from his source. For instance, in chapter 75.1, where a recipe for poisonous arrows is described, we read 'when I was looking for it and I was not able to find it, one of the most esteemed doctors gave me another drug of equal strength'.[57] This passage has in fact been copied in the first person, as the testimony of the *Apparatus Bellicus* suggests.[58] To this we can add another five cases, found in the first half of the treatise, where our author speaks in the first person but he copies Onasander. For example, when we read 'I cannot praise as much as blame the generals who destroy their own defences', or 'I think that this [tactic] which is risky or too daring and dangerous [is] not a product of good judgment and tactical knowledge, but of luck', as well as, 'I would allow [some] soldiers to run risks out of desire for distinction', we in fact have a copy of Onasander's view.[59] These, however, are minor cases which neither influence the content in any way nor contradict previous material. Therefore, not only did the content remain unaffected, but also the author underlined his erudition and directly connected himself with the credibility of his source.[60]

The only exception can be spotted in chapter 53.8, where the author of the *ST* has copied from *LT* (19.64). The *ST* reports that 'The so-called strepta, namely those which mechanically shoot the liquid fire (. . .) and the so-called hand-siphons, the very thing which Our Majesty presently invented, get the better of wooden towers'. This is significant because it strengthens the connection of the author of the *ST* with Leo VI since the latter wrote in his *Taktika* that 'These are also called hand siphons and have been fabricated recently by Our Majesty'.[61] The importance and meaning of this passage, however, are open to interpretation and depend on the dating and the authorship of the *ST*.[62]

Conclusion

The issue of the sources of the *ST* is indeed a complicated one, and it is partly responsible for the low reputation of the manual. Although the imitation of ancient and Byzantine sources was commonplace, the author of the *ST* is sometimes identified as a particularly slavish example: one who copied his material directly from two lost sources with little regard for consistency or for presenting the reality of his time. This assertion, however, not only contradicts the remarks of the author himself in the introduction,[63] but it is also contradicted by his working method regarding the treatment of his sources, which is actually free.

While the existence of the second lost source, the *Corpus Perditum*, seems to be more or less secure, the same does not apply to the first one, the *Tactica Perdita*. The existence of the latter seems unfounded when one studies the inconsistent cross-references more closely and recognizes the same free treatment of sources for the first and the second part of the work. Instead, the author of the *ST* seems to have used only one major lost source: the *Corpus Perditum*. For the other parts of the work, he could have used a small lost source or another version of known sources, but the majority of his work seems to derive from individual known and recognized sources, most of which are paraphrased.

Therefore, the author of the *ST* seems to have belonged to the norm with regard to his practices. He appears to be as credible as any other author of military manuals. Be that as it may, it would be worth turning to internal evidence to recognize the dating and the author of the text. It is also safe to argue that material and comments which are not found in parallel extant sources could well have been added by the author of the *ST*. These novelties can be taken into consideration to assist in the clarification of the issues discussed later.

Notes

1 Vári 1927: 265 (my trans. from the German).
2 Hunger 1969–1970: 22, 29–30; Nilsson 2006: 51–2; Nilsson and Scott 2007: 319–32.
3 Nilsson 2010: 196–8; Cutler 1995: 209–10.
4 Theon, 62.10–33 (trans. Kennedy, p. 6).
5 Hunger 1969–1970: 22, 24, 27–30; Lemerle 1971: 301–7; Kennedy 2003: ix–xiv; Odorico 2006: 213–15; Kaldellis 2007: 42–189; Webb 2009: 39–47; Nilsson 2010: 198.
6 Psellos, *Chronographia*, 6.61; Harris 2014: 22–3; Mango 1980: 147; Lemerle 1971: 255–7.
7 For a discussion of these aspects, see: Odorico 2006: 214–31; Nilsson 2010: 200.
8 Lemerle 1971: 255–7, suggests that around 300 pupils received an elite education in the tenth century. For education and literacy of the military aristocracy see: Haldon 1999: 270–4; McGeer 1995a: 138, 191–4; Browning 1978: 39–40, 42–4; Andriollo 2014: 131–8; Rance 2017a: 292–3. Among the books that Eustathios Boilas owned in 1059 were the Poems of George of Pisidia and probably the history of Agathias or Procopius' *Persian Wars*, see Vryonis 1957: 269–70.
9 Kekaumenos, 2; *TNO* 65.140–7; McGeer 1991: 138; Roueché 2002: 111–38; 2003: 23–37.
10 Psellos, *Chronographia,* 7.180; Anna Komnene, 15.3.6.
11 See Chapter 6, pp. 88–92.
12 See for example: Heron of Byzantium, *Parangelmata Poliorketika*, 3; Haldon 1990a: 70–4; Kolias 1993b: 39–44.
13 *ST,* 1.6; 1.27; 4.3. There are also cases when the Attic appears in our author's sources, see for example φυλάττειν in *ST,* 13.1, which also appears in Onasander, 38.7.
14 πυρὶ καὶ ξίφει, *ST,* 11.1,2, 23.6. See for instance: John Chrysostom, *Sanctum Pascha*, 47.9; *Omnes Sanctos*, 304.20; Gregorius, *Commentarius*, 6.12; John Damascenus, *Vita Barlaam*, 26.12; Euthymius, *Theodore Stratelates*, 8.17; *SEC, September,* 22.3.17.
15 Strabo, 12.3.18; Aeneas the Tactician, 33.3; Anna Komnene, 4.1.2.2; 4.4.6.2; 4.5.1.2; see also: Sullivan 2010a: 155.
16 See for example: *ST,* 6.13, 35.1–5, 38.3–8, 40.7, 44.4, 55.1, 54.9, 55.4. For the *allagia* in the tenth century see: Pertusi 1956: 92–5; Guilland 1967: 524–5; Haldon 1984: 275. *Allagion* did not only denote an army unit, it also meant exchange of prisoners mostly between Arabs and Byzantines, see for example, Kolia-Dermitzaki 2000: 608–12. For some of these terms, see also: Sullivan 2010a: 155.
17 A detailed, annotated and up-to-date list of the extant sources used by the author of *ST* can be found in Chatzelis and Harris 2017: 120–50.
18 Some of these sources have been identified by Dain 1946: 130–3 and Mecella 2009: 98–101.
19 For the dating of the *PS*, see: Dennis 1985: 2–4; Zuckerman 1990: 216; Cosentino 2000: 262–80; Rance 2007a: 719–37.
20 Vári 1927: 269.
21 *LT* 6.27–33; 4.58–63, 14.86–8.
22 *ST,* 2, 36; *LT* 4.1; 11.9, 20.209.
23 Haldon 2014: 45–7, 418.
24 *Poinalios Nomos*, 48.2; *Procheiros Nomos*, 40; Dain 1938: 8–9; 1946: 130–3; Dain and Foucault 1967: 357 believed it was the *Ekloga* which served as a source for chapter 50.4, but the wording and dating of the *Procheiros Nomos* dispute this view. See also Haldon 2014: 67; Mecella 2009: 100.
25 *ST,* 1.7, 1.10; Basil I, *Hortatory Chapters*, 1.11, 1.65.
26 For a comprehensive study of Byzantine *anthologia*, see Searby 2007: 1–112. For the use of such texts in rhetoric and hortatory works see, among others, Roueché 2003: 23–37; Markopoulos 1998: 469–79; Gerlach 2008: 599–600.
27 For more information, see Chapter 5, pp. 73–5.

28 See, for example, Stobaeus, 3.17.17; *Corpus Parisinum,* 3.453 as a source for *ST,* 1.7. Stobaeus, 4.10.29 as source for *ST,* 1.11. Stobaeus, 3.1.18; *Corpus Parisinum,* 6.66 as a source for *ST,* 1.14.
29 Isocrates, *To Nicocles,* 24 as a source for *ST,* 1.28.
30 Julian of Ascalon was an architect who most probably lived in the sixth century, see Hakim 2001: 4–25; Saliou 1996: 9–29, 79–132.
31 Dain 1938: 9 and Geiger 1992: 31–43 suggest that it belongs to Julian of Ascalon; c.f. Diller 1950: 22–5 and Saliou 1996: 21–7.
32 For the *Apparatus Bellicus,* see: Vieillefond 1932: xxxvi–xliii; 1970: 194, n.6; Zuckerman 1994: 359–89; Mecella 2009: 87–98; Wallraff et al. 2012: xlviii–lii; Rance 2017a: 324–38
33 Dain 1937: 73–86; Dain and Foucault 1967: 337; Schindler 1973: 205–16; Krentz and Wheeler 1994: xx–xiii. For stratagems in Byzantium, see also: Wheeler 1988.
34 *ST,* 56. The passage is also found in *MS,* 12D. Rance 2007b: 197, n.9, argued that the attribution to Urbicius was based on a misunderstanding, c.f. Dain and Foucault 1967: 341–2, 352–3, 372.
35 Rance 2017a: 347–56; c.f. Dain 1939: 53–4; Dain and Foucault 1967: 323, 339; Krentz and Wheeler 1994: xxii.
36 Vieillefond 1932: lii–liv; Dain 1938: 8–9; 1939: 70–1; Dain and Foucault 1967: 350–1, 353, 357–8.
37 Vieillefond 1970: 194–8; Mecella 2009: 107–13; Dain 1939: 14–31, 40–5; Dain and Foucault 1967: 353; Rance 2017a: 338–46.
38 Dain 1938: 8; Dain and Foucault 1967: 351.
39 These can be found in: *ST,* 21.4, 22.5, 39.3, 45.25, 45.26, 46.5, 49.8.
40 (trans. Chatzelis and Harris, p. 38).
41 *ST,* 21.5; 40.3; 43.10; 44.4; 46.23.
42 (trans. Chatzelis and Harris, p. 39).
43 (trans. Chatzelis and Harris, p. 55).
44 (trans. Chatzelis and Harris, p. 68).
45 (trans. Chatzelis and Harris, p. 83).
46 (trans. Chatzelis and Harris, p. 66).
47 The revision is discussed in Chapter 4, pp. 64–7.
48 The source for the *ST,* 53.9 is not Polybius, 21.28, as has been argued by Dain, but *Hypothesis,* 56.7, the wording of which is significantly more similar. See also Dain 1939: 58–9 and Mecella 2009: 100.
49 Dain 1937: 44, 55, 57–9, 81; 1938: 8; 1939: 69; 1946: 130–1; Dain and Foucault 1967: 357;Mecella 2009: 101.
50 See for example, *ST,* 16.3, with Onasander, 37.3; *ST,* 59.3, with *Apparatus Bellicus,* 2.6. These are discussed in Chapter 5, pp. 73–6.
51 For instance, the *PS,* 39.39–43, 13.61–71, 29.25–30 could be the source of chapters 48.5, 53.6 and 22.6. Chapter 2 seems to derive from *LT,* 11.9 and 20.209. Similarly, chapter 55.4 of the *ST* is similar to *MS,* 10.4.34–40 and *LT,* 15.62, while chapter 1.15 of our manual seems to come from *MS,* 8.2.77, 9.1.5–11 and *LT,* 29.124. Likewise, chapter 54.3–4 derives from *LT,* 15.27. Other examples include: *ST,* 20.5, derives from *MS,* 7.16B.20–37 and *LT,* 12,87–8, 14.30, 17.82–3. *LT,* 15.4,23 could be the source of *ST,* 54.1, some parallels are also found in *MS,* 9.3.75–81, 10.1.11–18. The information of chapter 55.2 could derive from *MS,* 10.4.41–63 and *LT,* 15.63–4. Chapter 52.1–2 of the *ST,* seems to have used *MS,* 10.2.1–14 and *LT,* 17.59–61 as sources. *LT,* 16.2–5, could be the source for *ST,* 50.1–3. For a more detailed record of parallel passages see Chatzelis and Harris 2017: 120–50.
52 Hunger 1969–1970: 15–17; Mango 1975: 16–18; *ODB*: ii.989. See also: Chapters 5 and 6.
53 (trans. Chatzelis and Harris, p. 91); (trans. Dennis, p. 379).
54 See also Bryennios, 4.5.2–5 with Neville 2012: 40, where Bryennios apologizes to his readers for being unfamiliar with the classical name of a river.

55 See Chapter 6, pp. 88–92.
56 Nilsson 2010: 199; Cutler 1995: 208–14; Lemerle 1971: 301–7. See Codoñer 2014b: 61–90 for a study of the strategy of this rewriting processes.
57 (trans. Chatzelis and Harris, p. 97).
58 *Apparatus Bellicus*, 37; Dain 1938: 8.
59 *ST,* 1.30, 1.36; 6; 10.1; 10.3 (trans. Chatzelis and Harris, pp. 25–7, 29, 31); Onasander, 32.
60 See also Chapter 6, pp. 88–92.
61 (trans. Chatzelis and Harris, p. 88); (trans. Dennis, p. 529).
62 See Chapters 3 and 4.
63 *ST,* 1.1.

3 The dating of the *Sylloge Tacticorum*

The *ST* is a very unusual case in that it gives very precise information on its date of composition and on attribution. The title of the treatise reads 'Λέοντος δεσπότου Ῥωμαίων αὐτοκράτορος ἔτους, ϛυιβ' [903–904]'.[1] Paradoxically, despite this unique evidence, the issues of authorship and dating can prove very problematic when studied together so that there is no scholarly consensus on either. An attempt to study simultaneously the issues of dating and attribution has very often led to a dead end and enhanced the confusion revolving around the *ST*.

The problem of the dating of the *Sylloge Tacticorum*

If one addresses the issue on good faith and chooses to accept that it was indeed Leo VI who compiled the *ST* around 903, he or she is ultimately faced with the task of identifying the relationship of the *ST* with *LT*.[2] A passage in *LT* could tempt us to argue that the *ST* was written before *LT*. Leo VI refers in his *Taktika* to a 'corresponding single volume of the Tactics', so one may envisage this as a citation to the *ST*. Save for this indication, however, this theory appears to be the most difficult to substantiate. The mere fact that the author of the *ST* has used *LT* as one of his sources is enough to undermine it severely, and Leo VI's reference to a volume of tactics cannot be conclusively linked to the *ST*, as it could very well refer to an unknown work or the treatise of Leo Katakylas, who, as we know, wrote a manual at the request of Leo VI.[3]

An approach that dates the *ST* after *LT* stands on more solid ground. Accepting that the manual was written in the reign of Leo VI but after *LT* would justify the use of *LT* as a source, given the similar military hierarchy that exists in both manuals, as well as the simultaneous reference to hand-siphons as a recent invention. Nevertheless, certain very important issues remain unaddressed. First, if both manuals were written at the time of Leo VI, how could they present such different material? For while there are many similarities between *LT* and the *ST*, the latter preserves a number of military innovations and an army which seems to have gradually evolved from that of *LT*.[4] These new material and tactics cannot be explained by a dating too close to *LT*, especially since the latter most probably dates to c. 904–912. This issue can be tackled to an extent if one argues that the *ST* was initially written in the reign of Leo VI but was later revised (c. 950) to include these new tactics and innovations like the *kataphraktoi* and the *menavlatoi*.[5] But this theory raises new issues, namely if the innovations of the *ST* were added later, why would Leo VI bother writing two separate treatises with such similar material in such a short period?

These imperative problems compel us to turn away from the reign of Leo VI and to consider a different dating than this which appears on the title of the treatise. The most popular alternative is to date the *ST* towards the middle of the tenth century. Paradoxically, while this theory appears in most books of general reference, there has never been a detailed study to provide extensive argument in support of it.[6] A dating around 950 can be very attractive at first. L preserves the *ST* along with medical and veterinary treatises attributed to Constantine VII, which may lead us to assume that the *ST* was also compiled at that time and that L was created to include works compiled in his reign.[7] This argument, however, is suspect on codicological grounds. We have already seen in the introduction that the *pinax* of the *ST* also lists the contents of the *Poinalios Stratiotikos Nomos* and the *Akolouthia*, which strongly implies that at an earlier stage the *ST* was part of a codex with only these two works, which were anyway compiled before Constantine VII.

Nevertheless, a theory which dates the *ST* around 950 is in accordance with scholarly tendencies that connect the appearance of specialized units, such as the *kataphraktoi* or the *menavlatoi* with the reign of Constantine VII. This approach is usually substantiated with al-Mutanabbī's description of Byzantine heavy cavalry in the battle of al-Hadath. However, this can end up being a circular argument, and one may point out correctly that since the *ST* is the earliest witness of these innovations, it could well mean that they had appeared earlier than 950.[8] Moreover, al-Mutanabbī' could have referred to the Byzantine *kataphraktoi* merely to highlight the victory of his patron, Sayf al-Dawla, rather than to record something unusual. It is, after all, a characteristic of Byzantine sources before 950 to present battles and campaigns without much detail; hardly any information about units and tactics is given before the *Vita Basilii*, the history of Leo the Deacon, and the appearance of promotional literature which described the deeds of generals in detail.[9] Therefore, the silence of the sources cannot be explicitly interpreted as proof that such developments did not take place earlier.

Likewise, due to the fact that this genre flourished during the reign of Constantine VII, it is also a trend to automatically link to his time most works which are compilations and excerpts.[10] Consequently, one may assume that the nature and the title of the manual itself, 'Compilation of Tactics', can be suitably placed in this chronological context. This connection, however, is by no means conclusive; it is well noted that this trend had already started in the ninth century, featuring, among others, legal corpuses and military manuals, such as the *Hypothesis* and the *Apparatus Bellicus*. The compilation of such treatises continued into the eleventh century with the *TNO*, which also contains extensive material from classical and older Byzantine authors that is arranged thematically.[11]

Therefore, a dating around 950 seems to be founded mostly on misconceptions and disproportions of focus. It is a non-conclusive theory which is the result of the fact that the second half of the tenth century has been studied disproportionately more than the first half. Sometimes things are readily attributed to the second half of the century, often ignoring that certain developments had taken place earlier than 950 and overlooking certain biases or methodological problems. For instance, modern scholarship often focuses more on Byzantine expansion after 950 and credits the reign of Constantine VII excessively for it. It is true that Theophanes Continuatus describes the Byzantine army at that time as bravely standing its ground and fighting against the Arabs without hiding, retreating or hesitating, but that is certainly a hyperbole.[12] For no other source records any significant change in the way the Byzantines fought. Moreover, we have already seen how Byzantium after 925 was gradually taking the initiative and shifting to a more offensive struggle against the Arabs, and other scholars have analysed how the policy of Constantine VII was, more or less, similar to that of Romanos I, save for more extensive annexations.[13] Thus, this praise for Constantine VII is something one would expect to find in Theophanes Continuatus since it was this emperor who commissioned and sponsored the writing of the history. As we have seen earlier with compilation literature, similar problems are not only found in warfare, but also in the cultural history of the tenth century.

Now that we have explored the problem of dating in some detail, we ought to turn to the text itself in order to determine whether we can find internal evidence which can help us date the *ST* with more confidence. Such evidence seems to be more or less credible, as we have established that the author of the *ST* did not carelessly copy from lost sources, and his material can be trusted as much as in other sources. The issue of the dating will be studied separately for now so as to remain uninfluenced by a potential false attribution, while the material of the *ST* will be approached critically and comparatively with the military, administrative and political milieu of the empire.[14]

A new dating for the *Sylloge Tacticorum*

Before we proceed to a new study of the internal evidence of the *ST*, it will be helpful to use Gilbert Dagron's criteria as our guideline. Dagron argued that the modernity of a military manual can be determined by focusing on three main factors: the military innovations and technology that are recorded, the attention and description given to the enemy and its tactics, and the relationship of the army with the administrative and socio-political context.[15] We will now explore these three criteria by first looking into evidence that is related to technology and innovations.

In his introduction the author of the *ST* promises to focus on and discuss contemporary armament and tactics. This not a mere *topos* in our case, since the *ST* is the earliest witness to a number of innovations which include new battle formations, new types of units and new usage of siege engines. While some of the innovations, like the *kataphraktoi* and the *menavlatoi*, do not conclusively clarify the problem of dating, the technology of siege warfare can prove more helpful. The *ST* is the first manual to record the use of Greek fire and hand-siphons for siege purposes, rather than just for naval warfare, as is the case with *LT*.[16] But while our author has made the effort to update the tradition, he is silent about another major invention, the *laisai*.

The *laisai* were a light type of tortoises, wooden shelters for covering troops during a siege.[17] These are first attested as having been employed by the Bulgarians against the Byzantines during the reign of Leo VI.[18] Some decades later, however, the Byzantines started to use them, and the treatise of Heron of Byzantium, the *Parangelmata Poliorketika*, is the first to record their use, commenting, in fact, that they were recently invented.[19] Could it be that the *ST* was written before the Byzantines started to use the *laisai*, or was it perhaps an unimportant development that the author did not feel the need to record?

It is certainly odd that the author of the *ST* did not mention the *laisai*, especially if we take into consideration that he had listed a number of other innovations which were not as significant. The importance of the *laisai* is highlighted by the fact that it is mentioned in every tactical manual dating after Heron, save for the *PM*, which does not have a chapter devoted to sieges, and the *DV*, whose chapter on sieges (21) does not refer specifically to any siege engine.[20] Therefore, it seems possible that the *laisai* were not employed by the Byzantines at the time the *ST* was compiled and that this manual was written before the *Parangelmata Poliorketika*.

If we accept this reasoning, then the dating of Heron of Byzantium is key to determining the dating of the *ST*. The material of the *Parangelmata* fits the context of the Byzantine offensive in the time of Romanos I and Constantine VII, as the words 'capture the cities of Agar' imply.[21] The earliest dating assigned to the *Parangelmata* is 934, but there is absolutely no evidence to support such a precise date.[22] Another theory dates the treatise to the sole reign of Constantine VII, and there is indeed good evidence to support that.[23] If we accept the second view, which is more secure, it would mean that the *ST* was probably written c. 920–950. What remains to be seen is to determine whether the other novelties of the *ST* are in line with such a dating.

Evidence from another source seems to imply that one of the innovations of the *ST* could date to the reign of Romanos I. In the ninth century, the *peltastai*, or the javelin men, were counted among the light infantry, together with the archers.[24] Accordingly, in the early tenth century we are informed from *LT* that the infantry of his time was only divided into heavy and light since the *peltastai* were still considered light infantry.[25] In the *ST*, however, that is not the case: the infantry is divided into three categories: heavy, medium and light. The medium infantry is called *peltastai* and is equipped almost as the heavy, save for the use of smaller shields, javelins and helmets which do not cover the face and for the absence of leg and arm guards. This medium infantry was drawn up and operated together with the heavy – the two together were called shield-bearing infantry. This development appears in all the later manuals of the tenth century, where the medium infantry is once again described in the same way but is designated with the term *akontistai*.[26]

This innovation can prove helpful for the dating of the *ST* if we link it with the information provided by Constantine VII in the *De Thematibus*. Constantine informs us that:

The so-called *tourmarchs* were appointed in the service of the *strategoi*. This rank signifies he who commands 500 archers, 300 *peltastai* and 100 heavy infantry. For this is how it is found in the book of John Philadelphos, the so-called Lydos.[27]

We read therefore that there was a triple division of the army of the Anatolikon *thema*. The *peltastai* were not included in the light infantry together with the archers, as is the case in *LT*. On the contrary, the *peltastai* appear as a distinct type of infantry between the archers and the heavy infantry, while Constantine VII uses the term *peltastai*, which is the same term that denotes the medium infantry in the *ST*. This passage is not without difficulties though. Constantine VII connects this information with John Lydos, who was active in the sixth century and who wrote the book *On the Magistracies of the Roman Constitution*. This information, however, is incompatible with the sixth century. First, a relevant citation cannot be found in John Lydos' book, and second, the term *tourmarches* was not in use during his time. On the other hand, this passage fits well into the known tenth-century context of *themata* and their manpower, which has led certain scholars to interpret it as a reflection of contemporary practice.[28]

Nevertheless, this is where we come to the second problem, for the dating of the first half of the *De Thematibus* is questionable to some extent. Some scholars believe that it was written in the reign of Romanos I because it contains praise to him, but others have noted that the reference to the transfer of the relics of St. Gregory from Cappadocia to Constantinople must place the work after 946, which is the date that the transfer took place.[29] But while this argument is very secure, this theory does not sufficiently explain the favourable comments for Romanos I. The reasoning that Constantine VII wanted to be politically correct because the text was intended for a wider readership is not very convincing, since Constantine VII did not seem to have had such problems when he abused the name of Alexander in *LT* and in a funeral poem dedicated to Leo VI.[30] If Constantine VII was not willing to do that for a relative, I do not see the reason for doing it for Romanos I.

That being said, it may be that the first half of the *De Thematibus* was indeed drafted when Romanos I was still alive. The work must have been revised in the later years of Constantine VII, but it seems that it remained unfinished, and therefore the praise was not edited.[31] In this light, it could be that Constantine VII is providing us with information which is contemporary to the reign of Romanos I, and this could also imply that the introduction of the *peltastai* as a distinct medium infantry unit took place during the reign of Romanos I.

The fact that the *ST* records the appearance of new battle formations and specialized units, such as the *peltastai*, the *menavlatoi* and the *kataphraktoi*, seems to be in line with the new challenges that the Byzantine army faced in the reign of Romanos I as it began to take the initiative against the Arabs. We have already seen that John Kourkouas participated for almost twenty years in offensive campaigns, which involved besieging fortresses and cities, as well as taking part in pitched battles.[32] While guerrilla tactics still remained important, this kind of warfare was different from the small defensive or retaliation operations which usually took place in the frontier. It required specialized infantry able to protect and assist the cavalry during marches, battles and sieges; specialized cavalry to take the offensive; solid battle formations; and a close cooperation of troops.[33] Therefore, all these developments in warfare could well have first taken place during the period 920–950 and then served as bases for further expansions and developments that lasted from the middle of the tenth century up to the beginning of the eleventh.[34]

This view is sufficiently summarized by Mark Whittow's words 'None of these appeared overnight (. . .) new armies and tactics had been developing since at least since 930'.[35] The aforementioned examples demonstrate that the military milieu during the period 920–950 was very fruitful for these innovations to have taken place and consequently for the drafting of the *ST*. It is, however, necessary to take the second criterion of Dagron into consideration and explore whether the *ST* presents the enemy in a way that it is compatible with this milieu.[36]

The *ST* does not contain extensive information about any specific enemy, but some useful conclusions can be drawn with the help of other sources. When all these are put together, it seems that the *ST* provides evidence which best fits into the context of Arab–Byzantine struggles of the first half of the tenth century. A first example can be seen in the absence of any mention of the *Arabitai*, who were nomadic Arab light cavalry troops. Their first incursions began in 902, but their earliest establishment did not take place until 936 near Aleppo. These nomads started to play a key role in the regions of Syria and Mesopotamia around 950, but their fullest impact was felt by the Byzantine armies after the death of Sayf al-Dawla (967).[37] Both *LT* and the *ST* are silent as to the threat posed by these men. In contrast, manuals which were written in the second half of the century, like the *PM* and the *TNO*, make explicit reference to the *Arabitai* and their tactics. These manuals also instruct the general to screen the flanks of the *kataphraktoi* wedge with the *prokoursatores* so as to ensure that the charge will be carried out without hindrance from the enemy light cavalry, a tactic which is not present in the *ST*.[38]

Another example which seems to paint a picture of the Arabs as they were in the first half of the tenth century comes indirectly from the *ST*'s information regarding the *menavlatoi*. The mere appearance of the *menavlatoi*, who were specialized infantry employed to repel the enemy heavy cavalry, seems to imply that there must have been a change in the Arab armies which made their use necessary. It was after the reforms of caliph al-Mu'tasim (833–842) that these developments started to take place, when specialized Turkish cavalry troops started to enter the service of the Abbasids. However, their impact was not that significant at first – it was not until the tenth century, and more specifically during the reign of al-Mu'tadid (892–902), that the system of *ghulams* was developed and appeared as a distinct elite force with separate command.

After 936, the *ghulams* fought in distinctive units and their numbers were a few hundred strong, while similar developments also took place in the autonomous Mesopotamia, where the main enemies of the Byzantines, the Ḥamdānids, started to employ them in the 930s as bodyguards in low numbers.[39] For example, during the reign of Romanos I, a force of 50 *ghulams* is reported, while at the battle of al-Hadath (954) 500 of them were said to have spear-headed the Arab assault.[40]

Therefore, from the tenth century onwards a steady increase in the number of heavy cavalry seems to have taken place in the Arab armies, while this type of cavalry started to play a more crucial role on the battlefield.[41]

The numbers and tactics of the *menavlatoi* in the *ST* seem to correspond to this stage of development when the numbers of enemy *ghulams* were still low. The maximum number of the *menavlatoi* was 300 in the *ST*, and they were expected to bear the heavy cavalry's onslaught on their own, distancing themselves forty *orgyai* from the infantry square formation.[42] This, however, is not the case in the second half of the century. In the *PM* the numbers of the *menavlatoi* are four times higher, while the same manual instructs the reader not to allow them to fight the enemy *ghulams* alone, but with the support of other infantry units.[43]

Similarly, the appearance of the *kataphraktoi* can be seen as a response to the deployment of the Turkish elite cavalry of the Abbasids. This mutual influence is supported by the similar equipment which the Byzantine *kataphraktoi* had with the Turkish or Arab elite heavy cavalry. Al-Mas'udi records that some Khazars and Arabs were heavily armed horse-archers also bearing lances and shields, while the *ST* instructs that the *kataphraktoi* should have shields, bows and lances as their primary weapons.[44] On the other hand, the *kataphraktoi* could have also appeared as an attempt to respond to developments in the enemy infantry. Some of these can be seen during the reign of al-Muqtadir (908–932) when the *Maṣāffi*, an elite infantry force, saw its numbers doubled and its pay dramatically increased during the period 917–929. Eventually their status was raised so much that it resulted in a civil war between the groups of infantry and cavalry.[45] The increase in payments would have most probably allowed the *Maṣāffi* to be equipped with the best equipment available, so the heavier the enemy infantry, the heavier the allied cavalry needed to break their formation. It is possible, however, that the decline of the caliphate and troubles in Baghdad may not have allowed the *Maṣāffi* to participate in frontier wars during the time of Romanos I, as has been argued that it was mostly local or regional troops that the Byzantines often faced.[46]

Be that as it may, part of the *Maṣāffi* was composed of Daylamite troops who not only served as palace guards but also as mercenaries employed in Syria and Mesopotamia. The Daylami, an enemy long known to the Byzantines, were specialized infantry who fought in thick formations, able to withstand cavalry onslaughts. Their numbers and importance, however, steadily increased in the tenth century. For example, the Abbasids regularly employed them after 929, while the same also applies to regional governors like the Ḥamdānids.[47] By taking this information into account, we can argue that the evidence of the *ST* regarding the enemy, either direct or indirect, seems to be compatible with the reign of Romanos I, not only because it corresponds to contemporary characteristics of the enemy but also because it justifies Byzantine developments as a response to them.

This moves us to the third criterion of Dagron and to whether the army of the *ST* is in accordance with the administrative and political milieu before 950. Our first evidence for a dating in the first half of the century comes from the study of the *taxiarchos*. The word occurs three times in the *ST*, where we read that 'it must be known that the *taxis* is also part of the formation, just like the *tagma*, and so the *tagmatarches* is also called *taxiarchos*'.[48] The *taxiarchos* therefore appears as an unofficial rank, used to describe an officer who commands a unit of drawn up men, regardless of whether this is an infantry or a cavalry unit. It is thus similar to the *tagma-tagmatarches*.[49] Consequently, this *taxiarchos* must not be confused with the official rank of *taxiarchos*, which appears in the second half of the tenth century and is specifically in command of an infantry unit, usually 1,000 strong, which is called *taxiarchia* and appears in the *PM* and *DRM*.[50]

Another evidence for a dating in the period 920–950 can be found when we look at the ranks of *droungarios* and *komes*. While the military hierarchy in *LT* and the *ST* seem identical, on closer inspection, all is not the same.[51] The author of the *ST* did not slavishly copy his information from *LT*, but he updated its information to make it relevant to his time. First of all, he reports that the cavalry units, which he calls *allagia*, consisted of a different number of men, while he also records that the *komes* could have commanded as few as fifty men, both of which are not found in *LT*.[52] Consequently, this seems to suggest that a small evolution had taken place in the meantime which would have been chronologically impossible to have been crystallized in the reign of Leo VI, given that *LT* was compiled quite late. This small evolution, however, certainly fits the context of the period 920–950, which can justify both a similar hierarchy with *LT* and a time frame that could allow such small changes to be developed and recorded.

With regard to the relationship between the rank of *droungarios* and *komes*, in the *ST* the *droungarios* is reported to have commanded from 3,000 to at least 1,000 men and the *komes* at least 50 and up to 400. These numbers also agree with the Arab geographer, Ibn Khurdādhbih, who reports that the *droungarioi* of the *thema* of Thrakesion commanded 1,000 men each and the *komites* 200. The same figures appear in Kudama for the 930s which are, however, most probably copied from Khurdādhbih.[53] What is of more interest to us now is not so much the exact number of men that a *droungarios* or a *komes* commanded, since those varied from *thema* to *thema* and from theory to practice, but the fact that the *komes* is clearly subordinate to the *droungarios* and generally commanded fewer men than him.[54]

Haldon has argued that already from the reign of Basil I (867–886) the clear distinction between the two ranks had gradually started to disappear. The *DC* records that in the ceremonial hair cutting of Basil's son, Leo, the 'θεματικοὶ ἄρχοντες τῶν δρουγγαροκομήτων' were among the officials who were present.[55] However, this seems to be the only reference which records the merging of the two ranks in the ninth century, and it is possible that either the author of the *DC* merged them together so as to generalize and record the presence of both or that he applied the situation of his time to Basil's. This becomes more evident from the *Kletorologion* of Philotheos (c. 899), where the *droungarioi* are stated to be first in the list of presence followed by the *komites*, a fact which confirms their higher hierarchy.[56] Nevertheless, as Haldon commented, the fact that they are referred to as 'δρουγγάριοι τῶν βάνδων, κόμητες ὁμοίως' seems to point towards the decline of their importance.[57]

The *DC* presents the *droungarios* and the *komes* as distinct ranks up to the reign of Romanos I,[58] but that is not the case for the sole reign of Constantine VII. The lists for the 949 expedition of Crete feature the joint rank of the *droungarokomites*, and the same also applies to the manual on imperial military expeditions.[59] These facts allow us to speculate that while this decline in the rank of *droungarios* had started by the middle of the ninth century, it became crystalized in the reign of Constantine VII.[60] Therefore, the fact that there is a clear distinction between *droungarios* and *komes* in the *ST* better fits the administrative context of the reign of Romanos I.

Similar conclusions can be drawn from other internal evidence of the *ST*. For example, our manual explicitly mentions two *tagmata* in the east: the royal *allagia* of Thrakesion and Charsianon.[61] The royal *allagion* of Thrakesion can correspond with the *peratika tagmata* which were stationed in Asia Minor, but not too far from the capital.[62] Troops from Thrakesion are reported to have fought against the Rus in 941 – perhaps they can be identified as the *tagmata* of Thrakesion.[63] The royal *allagion* of Charsianon remains a mystery. Our first clear mention of *tagmata* in the Charsianon dates to the eleventh century; therefore, the *ST* is the earliest source referring to them in the tenth.[64] These troops could have been recruited to the *tagmata* especially for the needs of a particular campaign, much like the Armenians of Platanion in the *DC*, or as *peratika tagmata* stationed there against the Arabs.[65] The fact that they were from Charsianon might not be a coincidence, as they could have been recruited from this *thema*, which was home to the powerful family of Argyroi.[66] The Argyroi were allied to Romanos I, as the latter married his daughter Agatha to the son of Leo Argyros. In addition, both Leo and his brother, Pothos Argyros, were *domestikoi* of the *scholai* before John Kourkouas, when Romanos I was in power.[67]

The recruitment of men from a loyal *thema* to the *tagmata* could be explained in the tenth-century context of powerful families securing a strong imperial force to use as a counterweight both against other strong families of this *thema*, like the Phokades, and generally against rebellions aimed at the throne.[68] For example, Skylitzes informs us that John Kourkouas confronted the rebellion of Bardas Boilas, *strategos* of Chaldia, Tatzates and Adrian Chaldos, when he happened to be at Charsianon. Kourkouas could have used the *tagmata* there to suppress the rebellion since he was *domestikos* of the *scholai* at that time.[69] Moreover, the recruitment of allied powerful families to the *tagmata* gave them the chance to enhance their position by acquiring booty, lands and fame by participating in the frequent operations of the nearby front, not to mention better payment.[70]

The *ST* also mentions the western *tagmata*, which can be identified with the *tagmata* of Thrace and Macedonia.[71] It is also possible, however, that some or all of the *allagia* mentioned in our manual were detachments of *tagmata* listed as having participated in a recent major campaign. Perhaps they were detachments of the imperial army who participated in the most important operation of that time, the capture of Melitene, since the chroniclers clearly inform us that both *themata* and *tagmata* participated in that campaign.[72]

Conclusion

When we take all of these factors into consideration, it seems that the date which appears on the title of the treatise cannot be trusted. The *ST* cannot be dated to the reign of Leo VI because it has used his *Taktika* as a source, and despite the many similarities, it presents a number of developments which cannot be contemporary with Leo VI. A dating in the reign of Leo VI remains problematic even if we suppose that the innovative material of the *ST* was added later, because in this case it is hard to see the purpose for the production of the manual in the first place. Likewise, a dating around 950 does not seem plausible, as it ignores the manuscript tradition and the internal evidence of the manual itself.

The *ST* presents an army with technological and administrative elements dating to the first half of the century and an enemy which seems in line with this dating. It seems that the most suitable dating for the *ST* is the reign of Romanos I Lekapenos (920–944). This is supported by all three factors proposed by Dagron to determine the modernity of a treatise. In terms of technology and innovations, the appearance of the *peltastai* as medium infantry could have taken place at the same time, while the *laisai*, which appears in all later manuals, is an innovation not recorded in the *ST*, despite the fact that the author presents us with an updated version of the Byzantine army. With regard to information concerning the enemy, the *ST*, contrary to later manuals, does not refer to the threat of nomad Arabs which gradually started to menace the Byzantines from 950 onwards, while the tactics of the *menavlatoi* imply a low number of enemy heavy cavalry, which agrees with developments that took place in the Arab armies at the first half of the century. Administrative information about the army in the *ST* also supports a dating at this period since there is a gradual development from *LT*, but the ranks of *droungarios* and *komes* are still distinct and not joined, as is the case in the sole reign of Constantine VII.

In addition, the reign of Romanos I seems very suitable not only because the Byzantine offensive brought new challenges which are in line with the innovations presented in the *ST* but also because these innovations can be explained as responses to contemporary Arab developments. In addition, the reign of Romanos I is the most attractive for the compilation of the *ST* because it was long enough to allow for these developments to take place, crystallize and be recorded, and yet was not too far from the context of the reign of Leo VI, from which they gradually evolved. Last but not least, this was a period when very experienced generals like John Kourkouas and Melias flourished and held posts for an extensive amount of time and one of the few times that the emperor himself was a man with military experience who had held the rank of *droungarios* of the fleet before he became emperor.

Although the internal evidence of the *ST* agrees with this new dating, it is also essential to examine this dating theory in connection with the authorship and the attribution of the manual. The discussion which follows in the next chapter reinforces in fact a dating at the time of Romanos I.

Notes

1 This is the title of the table of contents. The chapter heading of the first chapter is similar and reads: 'Ἐκ τῶν τακτικῶν Λέοντος δεσπότου Ῥωμαίων αὐτοκράτορος προοίμιον ἐν ᾧ καὶ ποταπόν δεῖ τον στρατηγόν εἶναι καὶ πόσα εἰσί τά τοῦτον χαρακτηρίζοντα ἔτους, ϛυιβ'.

2 See the introduction earlier for a detailed overview of previous dating theories, pp. 5–8.

3 *LT*, 2.33 (trans. Dennis, p. 37); Constantine VII, *Three Treatises*, C.24–39; Haldon 1990a: 45–53; 2014: 134. See Chapter 2 for a more detailed discussion on the issue of the sources.

4 See Chapter 5 for more information, pp. 78–83.

5 Haldon 2014: 55–68, 337–8, 359, 360.

6 See for example *ODB*: iii. 1980; Haldon 1999: 220–2; Parani 2003: 101.

7 McGeer 1995a.

8 Kolias 1993a: 24–6, n.10.

9 See Kolias 1993a: 11–36; Markopoulos 2009: 697–715 and for more details Chapter 6, pp. 97–100.

10 For a discussion of that trend and whether it can be termed as encyclopaedism see: Odorico 1990: 1–21; Flusin 2002: 556; Roberto 2009: 73–84; Holmes 2010: 56–60; Schreiner 2011: 3–8, 11–17, 23–5.

11 Németh 2010: 20–31; Dain 1953: 64–8, 71, 75, 79–80; Lemerle 1971: 242–4; Trombley 1997: 261–74; Roberto 2009: 71–3; Holmes 2010: 56. Other military handbooks, like *LT*, have similar characteristics: the main material is organized in relevant chapters and the last part of the work contains 'Various Sayings' usually deriving from older sources and sometimes repeating what has already been said in the first part.

12 *TC*, 459–60.

13 See for instance Shepard 2001: 19–40.

14 The issue of authorship and attribution will be discussed in Chapter 4, pp. 57–64.

15 Dagron and Mihăescu 1986: 142.

16 See Chapter 5, pp. 82–3.

17 While the employment of wooden hunt/fence-like barriers, goes back to the Roman times, known as *vinea*, their widespread used and the term *laisai/lesai* appears in the tenth century, see: McGeer 1991: 136; Sullivan 2000: 175–6.

18 *DAI*, 51.114–20; McGeer 1991: 136.

19 Heron of Byzantium, *Parangelmata Poliorketika*, 2.1–4.

20 In the *De Obsidione Toleranda*, 50.6, its dating is not certain, but it was written after 924. However, it does not refer to the *laisai* as recently invented, which may imply that it was completed after the *Parangelmata*. Dain 1940a: 136, proposed a dating around the middle of the tenth century. For further discussion about its dating see: Dain and Foucault 1967: 349–50, 359; van den Berg 1947: 3; Sullivan 2003a: 139–41. In the *Mémorandum*, 6, Dain proposed a dating at the second half of the tenth century, 1940a: 136; Dain and Foucault 1967: 366–7; Sullivan 2003a: 140. In the *DRM*, 27.7, see Dennis 1985: 241–3 and Dagron and Mihăescu 1986: 171–5 for the dating. In the *TNO*, 65.86–100.

21 Heron of Byzantium, *Parangelmata Poliorketika*, 58.6–10 (trans. Sullivan, p. 113); Dain 1933: 16–17; Sullivan 2000: 3–4.

22 Martin 1854: 275–7.

23 Schneider 1908: 84–5, proposed that the *Parangelmata* was connected with Constantine VII. Dain 1933: 16–17, at first doubted this theory, but later accepted this connection, stating that his various nominal references to his sources point towards this direction, see: Dain 1953: 77–8. Sullivan 2000: 4, 15–21, 248, noticed that the word θεολέστων used in this treatise (ch. 58.9) is employed in the works of Constantine VII, such as in the *DC* and his military orations. He also pointed how relevant the manual

is to a mid-tenth-century context. To those arguments we can add the author's manifesto on the avoidance of the Attic language and the use of simple flat writing with clarity (ch. 3).

24 In *Synagoge*, π.282, we read: πελτασταί = τοξόται, ἢ τοὺς ξυστοὺς κατέχοντες, while in the *Lexicon* of Photios π.408.12–13, we read the same with the addition that the πέλτη δὲ εἶδος ἀσπίδος οὐκ ἐχούσης ἴτυν οὐδ᾽ ἐπίχαλκον, οὐδὲ βοός ἀλλ᾽ αἰγὸς δέρματι περιτεταμένη.

25 *LT*, 6.20.

26 *ST*, 38.6–7; *DV*. 3.3, 20.73; *DRM*. 1.10–20, 35, 5.12–18, 6.1–5, 58–60; *PM*, 1.51–62, 82–7, 95–7; Dagron and Mihăescu 1986: 192, n.42; McGeer 1995a: 208–9. See also Chapter 5, pp. 78–9.

27 Constantine VII, *De Thematibus*, 1.1.73–7.

28 Pertusi 1952: 117; Ahrweiler 1960: 3; Cheynet 1995: 321; Treadgold 1992: 89–110; Haldon 2000: 305–29; 2014: 135–7. For the Anatolikon *thema* see: *TIB*: ii and iv.

29 Ostrogorsky 1953: 38; Huxley 1980: 31–2; Lemerle 1971: 279–80; c.f. Ahrweiler 1981: 1–5.

30 See Chapter 4, pp. 60–1.

31 Németh 2010: 51–2.

32 See Chapter 1, pp. 16–21. For an example of a pitched battle see: Vasiliev 1935–1968: ii.i.121–2.

33 See Chapter 5, pp. 78–82.

34 Runciman 1929: 137–50; Vasiliev 1935–1968: ii.257–300; Canard 1951: 731–51; Guilland 1950: 30; McGeer 1995a: 262; Haldon 1999: 218–22.

35 Whittow 1996: 323–4.

36 For Dagron's criteria see p. 42 above.

37 Von Sievers 1979: 230–4; Cappel 1994: 114–15, 129, n.4.

38 *ST*, 46.9; *PM*, 2.104, 128, 129, 4.126–37, 180, 187; *TNO*, 57.125–8, 167–59, 61.155–9, 256–7; McGeer 1995a: 68, 305; Cappel 1994: 114, n.2, 117.

39 Bikhazi 1981: 256–7, 372; Kennedy 2004: 204, 269; Gordon 2001: 127, 139. For more information on al-Mu'tasim's reforms see Ayalon 1994: 23–31; Gordon 2001: 119–40.

40 McGeer 1995a: 237; Vasiliev 1935–1968: ii.352–3, 362, ii.i.361–2; Canard 1951: 779–81, 794.

41 Bosworth 1965–1966: 153–4; Nicolle 1980: 10; 1983: 47–52; McGeer 1995a: 237.

42 *ST*, 47.16.

43 McGeer 1995a: 273–5. See also Chapter 7, pp. 158–60.

44 Mas'udi, 22; *ST*, 39.1–6.

45 Miskawaihi, 227; Kennedy 2001: 161

46 Whittow 1996: 329.

47 Vasiliev 1935–1968: ii.i.349; Bosworth 1965–1966: 148–51; McGeer 1995a: 234–5; Bikhazi 1981: 421; Kennedy 2004: 269–71.

48 *ST*, 22.8, 35.10, 50.6 (trans. Chatzelis and Harris, p. 51)

49 *ST*, 43.5, 46.25, 43.2, 46.22. For a similar case see: *PS*, 15.65–6.

50 *PM*, 1.75, 81, 5.21; *DRM*, 1.11–16; Oikonomides 1972: 273, 335–6; Kühn 1991: 273–8; McGeer 1988: 135–45; 1995a: 265–72; Haldon 1999: 218.

51 C.f. Vári 1927: 267.

52 *ST*, 35.2–5; c.f. *LT*, 4.47, 18.146.

53 Treadgold 1992: 88–90; Haldon 1999: 108; 2000: 322. For the *thema* of Thrakesion see: *TIB*: vii.45, 49f., 84f., 90f., 94f., A. 410.

54 See the detailed study of Haldon 2000: 305–30; c.f. Treadgold 1992: 84–110.

55 *DC*, 622.10; Haldon 2000: 324–5. For this ritual see Tougher 1997: 46–8.

56 *Kletorologion*, 109.23–4, 157.9–11.

57 Haldon 2000: 325.

58 For the expedition of Crete in 911, the 5,000 Mardaites included 44 *droungarioi* and 44 *komites*. The fact that they were both 44 implies a low commanding number of men for the *droungarioi*. The latter, however, received twelve *nomismata*, whereas the *komites* only six. The Armenians of Sebasteia numbering 960 men for the same expedition included 10 *droungarioi* and 8 *komites*; once more the *droungarioi* were paid more, namely six *nomismata* each, while the *komites* received five. See: *DC*, 656.10–15; Haldon 2000: 325; c.f. Treadgold 1992: 118–19.

59 The force of the *thema* of Thrakesion had sixty-four *droungarokomites* in 949. The situation of Charpezikion seems to be quite different – the lowest rank reported is the *droungarioi*, which probably implies that the *droungarioi* became equivalent to the *komites*, see: *DC*, 663.6, 667.7–11; Haldon 2000: 326–7; c.f. Treadgold 1992: 127–30. In the treatise on imperial expeditions the hierarchy is reported as 'All the strategoi should issue orders to their *tourmarchai*, and the latter to their *droungarokomites*, so that each and every *bandon* has its smith and likewise its bootmakers', see: Constantine VII, *Three Treatises*, C.653–5 (trans. Haldon, pp. 136–7); Haldon 1990a: 256–8; Kühn 1991: 51–2.

60 Haldon 2000: 328.

61 *ST*, 35.5. For the term in the tenth century see: Pertusi 1956: 92–5; Guilland 1967: 524–5; Haldon 1984: 275. For the Charsianon *thema* see: *TIB:* ii.

62 For *tagmata* stationed outside the capital see: Ahrweiler 1960: 25–33, 55–9; Guilland 1967: 428–30; Oikonomides 1972: 329–35; Toynbee 1973: 286; Haldon 1984: 234; 1999: 84; 2000: 332; Kühn 1991: 69; McGeer 1995a: 201.

63 The chronicles report that Bardas Phokas fought the Rus with some picked men but that Kourkouas was in charge of the main army that arrived later. The *Life of Saint Basil the Younger* states that Bardas was in charge of troops from Macedonia, and Spongarios from Thrakesion. Since Spongarios replaces the name of Theophanes who was in charge of the fleet, and Kourkouas had already with him the largest part of the Eastern army, it is possible that Bardas actually commanded some remaining detachments of the *tagmata*. See: Symeon Magister, 136.71–5; TC, 423–5; Skylitzes, 229; *Basil Younger*, 3.23–7. For Skylitzes' reference to *tagmata* in this passage see Holmes 2005: 146, who does not regard it as trustworthy.

64 Ahrweiler 1960: 34–5; Kühn 1991: 265.

65 *DC*, 652.6–7, 657.20–658.8, 666; Haldon 1984: 219–20; Haldon 2000: 333. Caesarea in Charsianon was an important site, being one of the imperial *aplekta* and probably also producing arms, see: Constantine VII, *Three Treatises*, A.11–12; Haldon and Kennedy 1980: 85–7; Huxley 1975: 90–3; Haldon 1984: 318–19; 1990a: 63, n.53; Cooper and Decker 2012: 242.

66 Recruiting thematic elites in the *tagmata* see: *DC*, 657.20–658.8; Haldon 1984: 219–20; Kühn 1991: 251–2; Grigoriou-Ioannidou 1993: 35–41. For the Argyroi family see: Vannier 1975: 15–17; Cooper and Decker 2012: 234–6, 250.

67 Guilland 1950: 28–9; Cheynet and Vannier 2003: 60–2; Kühn 1991: 79–80.

68 For example, the strong family of the Phokades was pushed to the background by Romanos but returned to the command of high-offices during the sole reign of Constantine VII, see: TC, 436; Skylitzes, 288; Runciman 1929: 64–6; Cheynet 1986a: 296–301; 1991: 210–12; Cooper and Decker 2012: 233–9, 250. For the powerful provincial families and imperial authority see: Lilie 1993: 66–7, 71–3; Howard-Johnston 1995: 78–89; Haldon 1999: 238–9, 273–4; 2008: 165–9, 172–6; Cheynet 2006: 31–2, 37–8; Stephenson 2010: 22–4; Andriollo 2017. For the political aspect of the *tagmata* in enforcing the central imperial policy see Haldon 1984: 231–5; Kühn 1991: 48–9. Numerous rebellions took place in the reign of Romanos I, some of the important participants involved were Leo Phokas and the *strategos* of Chaldia Bardas Boilas: Symeon Magister, 136.4, 26; Skylitzes, 210, 217; TC, 395–6, 404.

69 Skylitzes, 217. Lounghis 1997: 101, argues that we can interpret from this reference that Kourkouas used Caesarea as a temporary base for his troops. For the *thema* of Chaldia see: *TIB:* ix. 73f., 81, 86, A. 109.
70 Haldon and Kennedy 1980: 98–9; Haldon 2009: 185–96; Magdalino 2009: 228; Cooper and Decker 2012: 251–2.
71 *DC*, 652.4, 660.19, 666.
72 Symeon Magister, 136.53; TC, 416.

4 The authorship, attribution and redaction of the *Sylloge Tacticorum*

The previous chapter argued that the internal evidence of the *ST* fits the technological, military and administrative context of the first half of the tenth century, and more specifically that of the reign of Romanos I. To make the best of this reasoning and to better understand the manual itself, it is essential to study the dating in relationship with authorship and attribution. Therefore, we will attempt to investigate who was behind the compilation of the *ST* and to explain why it was falsely attributed to Leo VI. Hopefully, this study will not only clarify some very controversial issues of the *ST* but will also reinforce the new dating theory, shed some light on possible previous versions of the text and put all of them into context. In order to achieve this and to appreciate the matters that will be discussed, we will begin by looking into the problems which pertain to the authorship of the text.

The problem of the authorship of the *Sylloge Tacticorum*

Inevitably, the issue of the authorship of the *ST* cannot be seen individually from the dating of the text and from whether one chooses to trust the attribution in the title, as well as the internal evidence of the treatise. On the one hand are those who take the information in the title at face value and accept that the *ST* was written by an emperor, Leo VI. On the other hand are those who believe that the issue cannot be clarified because they distrust the information of the *ST* altogether, perceiving the latter as a product of careless copying from lost sources whose original material is impossible to be discerned.[1]

Our earliest observations, however, demonstrated that the author of the *ST* was not an extraordinary case of slavish imitation and that the bad reputation of the manual seems unfounded. In this light, the material of the *ST* is as credible as any other source of this kind, and it certainly allows for a more careful consideration which will enable us to examine the identity or rank of its author. Was the author of the *ST* an emperor as the title wants us to believe?

The imperial authorship of the *Sylloge Tacticorum*

To begin with, the fact that the treatise is attributed to an emperor may not be wholly misleading. For the content of the *ST* seems to be in line with an imperial attribution. The person who compiled the *ST* seems to have had access to the imperial library, not only because he had consulted a plethora of different sources but also because he used works which were not intended for wide readership, the most characteristic example being the *Hortatory Chapters* of Basil I.

Another indication is that the author of the *ST* seems to present himself as an emperor through the phrase 'our Majesty'.[2] This phrase may link the treatise with an imperial milieu since it is frequently used in texts commissioned by emperors to indirectly refer to their dignity as supreme rulers. Some tenth-century examples include the novels and *Taktika* of Leo VI and the *DAI* and the *DC* of Constantine VII, as well as imperial correspondence in general.[3] This evidence is not conclusive though, since there are cases in which this phrase is used to broadly refer to the dignity of the emperor and it is not connected with any particular figure. The most relevant example comes from the *Taktika* of Ouranos, who was a general and governor of Antioch. In the *TNO* we read 'They [the *prokoursatores*] must have one head commander either a *strategos* or someone else whom Our Majesty appoints'.[4] Therefore, Ouranos uses the phrase without being an emperor and without making a specific reference to Basil II; he just describes a responsibility that all Byzantine emperors had.

On the other hand, the phrase 'our Majesty' is by no means the only evidence which implies an imperial authorship or commission. When the author of the *ST* addresses the general, he does so in the imperative mood or through the use of the phrase 'the general must', which means that he outranks him, commanding him to act in a certain way.[5] These phrases occur numerous times in the text, and they cannot solely be attributed to dependence on older models, because they also appear in parts of the work which are original and innovative. The fact that the same tone is retained throughout the treatise, both in the first and in the second half, facilitates acceptance that it is not a product of copying the mood and style of sources, either directly or indirectly. For example, in chapter 59.3 the author of the *ST* states that 'we compiled this book judging that these [stratagems] and others of the kind should be recorded (. . .) so that our generals may be able to guard against them'.[6] The words 'our generals' imply that the author is to be distinguished from common generals and that he commands or 'owns' them.[7] After all, the use of such authoritative phrases and tone is common in manuals written or commissioned by emperors such as *MS*, *LT*, the *PM* and the *DV*.[8]

The authoritative status of the author becomes more obvious in chapter 54.1, where we read that 'The general (. . .) must first secure the camp in every possible way, which in chapter 22 Our Majesty ordered to be established standing off from the city at a distance of approximately two miles'.[9] This passage cannot be traced to any extant source, and the author clearly states that he considers the information provided in his manual to be an order. This order is either given by the author, who is the emperor and addresses himself indirectly, or by a ghost author, who writes in the name of the emperor.

Consequently, evidence throughout the text seems to suggest that the author of the *ST* was somebody who possessed very high authority. He was either an emperor himself or somebody who wrote on behalf of an emperor. As is always the case with these works, it is difficult to say whether the manual was personally written by an emperor or by a team of ghost authors under his auspices. It is common for military manuals commissioned or written by an emperor to use both the first personal singular and the first person plural, without necessarily denoting two different subjects. If one tries to speculate on such a distinction, it is usually very difficult, if not impossible, to recognize the difference.[10] The *ST* is not an exception; the 'we' and 'I' are used interchangeably in the text, sometimes mimicking its sources, sometimes not.[11] The only difference can be found in chapter 59.3 where the change from 'we' to 'I' might perhaps have some significance. The passage cannot be traced to any extant source, and it reads 'We compiled this book judging that these [stratagems] and others of the kind should be recorded not in order to be used by us against the enemy (for I believe that they are unworthy even to be mentioned in a Christian context)'.[12] Here the 'we' seems to be differentiated from the 'I', as the 'we' seems to play the role of a team who helped the 'I' in the compilation and collection of the material, but eventually it is the 'I', the emperor-supervisor, who expresses his personal view and dominates. Whatever the case, it is not unlikely that the author had a group of people from his court that he supervised and who probably did most of the work. What is more important now is to attempt to identify who this emperor was and to estimate whether this identification agrees with the chronical context of the new dating theory.

The emperor who commissioned the *Sylloge Tacticorum*

Trying to identify the emperor responsible for the *ST* can be a very tricky task. For the time being we will not take into consideration our findings regarding dating, as the fact that the contents of the *ST* best fit into the reign of Romanos I does not necessarily mean that the manual was indeed produced under his auspices. It could have been the case that it was produced by another emperor who failed to update its material into his own time, resulting in the compilation of a slightly outdated manual. Consequently, our *terminus post quem* will be the year 963, which is when Nikephoros II Phokas became emperor, since we know that he had used the *ST* as a source for the drafting of his *PM*.[13]

Although the title of the *ST* cannot be taken at face value, the fact that the content and tone of the treatise suggest an imperial commission allows us to explore the possibility that Leo VI was indeed its author. On closer scrutiny, however, and after one turns to other known texts commissioned by emperors in the tenth century to identify their trends and style, the theory crumples. The style and sources of the *ST* are very different from that of *LT*, so Leo VI cannot have been its author or commissioner. A comparative reading of *LT* and the *ST* is enough to establish that the author of the *ST* had his own unique style, and his work looks nothing like that of Leo.[14] The next candidate to consider is Alexander, the brother of Leo VI. While this theory solves the problem of different styles, it lacks cogency.[15] Apart from the fact that we know very little of his literary interests and that his reign was very short, Alexander was a contemporary of Leo VI and such a difference and evolution in tactics between the *LT* and *ST* cannot be explained.

On the other hand, the commissioning of the *ST* by Constantine VII is very attractive at first sight. Constantine had definitely shown an interest in the compilation of new treatises or in the copying of older ones, either classical or Byzantine. For instance, he was responsible for the *DAI*, the *DC* and the three treatises on imperial military expeditions, and he also seems to have been behind the production of the *Laurentianus Plut.* 55.4, a manuscript which contains a number of military treatises, such as the works of Aelian, Asclepiodotus, Onasander, the *MS*, the *Hypothesis*, the *PS* and *LT*.[16] The manuscript tradition of the *ST*, however, implies that the treatise must have been written before the reign of Constantine VII because the manual first belonged to a codex with the military laws of the *Poinalios Stratiotikos Nomos* and a military hymn called the *Akolouthia*, which both seem to have been contemporary with the reign of Leo VI.

Yet this evidence cannot exclude for sure the possibility that Constantine VII was the emperor who commissioned the *ST*. To do justice to this theory, one has to turn to a comparative reading of works which are connected with Constantine VII in order to determine whether their style or their general characteristics match those of the *ST*. This endeavour can prove very problematic though, because works such as the *DC*, the *DAI*, the *Vita Basilii*, the *De Thematibus* and the *Geoponika* are themselves very different and belong to different genres or sub-genres. What is more, the extent to which Constantine VII was actually involved in the writing procedure or what exactly his supervision and his relationship with his team and ghost authors involved were are also unclear.[17]

It is probably because of these problems that some common characteristics which appear both in the *ST* and in works connected with Constantine VII cannot provide us with conclusive evidence. For these similarities appear to be more generic and cannot be attributed solely to Constantine VII. One such common feature is references to Solomon. The *ST* records that if the general 'does not boast the wisdom and judgment of Solomon (. . .) he will not be able to accomplish anything beneficial'.[18] Solomon also plays an important role in the works of Constantine VII.[19] For example, in the three treatises of imperial military expeditions, Constantine introduces himself as Solomon to his son Romanos II, while in the *DC* his throne is called the throne of Solomon.[20] But this symbolism cannot explicitly link to Constantine VII, as it was a more general phenomenon in the Byzantine world. Other members of the Macedonian dynasty, such as Leo VI, were also compared to Solomon, and to make matters worse, the name of Solomon also features in an oration which celebrates Romanos I's peace treaty with the Bulgarians in 927, as well as in Romanos I's imperial correspondence with Symeon I.[21]

Another aspect of the known works of Constantine VII is etymological comments which are given usually with the words 'in the Roman language'.[22] Sometimes the author of the *ST* does provide us with alternative terms using a similar phrasing. For example, we read that 'at this time the *skopoi* as the Greeks call them, or the so-called *viglai* in the Roman language, should also be dispatched' or that 'the depth or thickness of each *tagma*, which is also called *kontoubernion* in the Roman language, comprises of sixteen ranks'.[23] Once again, however, these are found in Byzantine treatises in general and in a very similar fashion, as is the case in *LT*, where we read that 'those before us, the more recent tactical writers, called [the exhorters] by the Latin term *cantatores*' and that 'the army will proceed more safely along the road (. . .) if you reconnoitre the paths that lie ahead of you by sending out a few men who are called *minsoratores* and *antikersores* in the Roman tongue'.[24]

Similar observations for the possible authorship of Constantine VII can be made with regard to the aforementioned comment of the author of the *ST* in chapter 59.3. After presenting material which originally derives from Julius Africanus on how to deal with the enemy using disease and poison, the author comments that these are unworthy even to be mentioned in a Christian context, explaining that these details were added to the *ST* merely as a precaution for Byzantine generals. Comments of the same kind can be spotted in the *De Thematibus* and the *Geoponika* following passages which cite works from antiquity; for instance,

> So much for what the ancient sources say. I think that some of these methods are quite improper and to be avoided; I advise readers not to trust them completely. I copied them so as not to be thought to have omitted anything said by the ancient sources.[25]
>
> The so-called Aegean Sea is also considered a *thema*. According to those who make commentaries on Homer, it obtained its name from Aegeus the son of Poseidon. But Greek history is not serviceable now, for it is full of fallacies.[26]

These comments therefore are a general characteristic of Byzantine authors and of the way they used and admired the classical tradition without fully accepting it, since it reflected a pagan culture.[27] To demonstrate that with another tenth-century example, after Leo the Deacon described Aristotle's theory of how earthquakes occur, he then commented that 'the foolish babbling of the Greeks has explained these things the way they want it; but I would go along with the holy David and say that it is through the agency of God that such quakes happen to us'.[28]

The final putative link to Constantine VII can be seen in the use of the phrase Ἰστέον ὅτι', which is found numerous times in works which he was involved with, notably the three treatises on imperial military expeditions, the *DAI* and the *DC*.[29] In the *ST* we most probably find its abbreviated form when various chapter headings begin with the word 'ὅτι'; for instance, 'Ὅτι τὰ ὑπεσχημένα τοῖς προδόταις ἀπαράθραυστα χρὴ τηρεῖν' or 'Ὅτι δεῖ κρύπτειν ἐν τῷ στρατοπέδῳ τὰς ἰδίας συμφοράς'.[30] Despite the fact that some scholars regard this phrase as a characteristic of a Constantinian involvement, this is by no means conclusive.[31] The phrase also appears in works that have nothing to do with him. Among them are a small treatise on siege warfare called *Mémorandum* and the manual of harmonics of Nicomachus the Pythagorean (c. 100 AD).[32]

In themselves these features do not provide enough evidence either to accept or to discard Constantine VII as the commissioner of the *ST*. Fortunately, more conclusive evidence can be found in the way in which the Macedonian dynasty highlighted its legitimacy. Denoting imperial legitimacy and dynastic continuity through literary works was a characteristic of the Macedonian dynasty already from the late ninth century, and Constantine VII upheld this tradition.[33] For example, Leo VI refers twice to his father, Basil I, in *LT*, and Constantine VII does so in the *De Thematibus*, *DAI* and *DC*.[34] Accordingly, Constantine VII also mentioned his father, Leo VI, quite frequently in his works.[35]

In contrast, such references are completely absent from the *ST*, and the name of Leo VI and Basil I is not commemorated even once. What is even more puzzling though is that the author of the *ST* seems to have deliberately decided not to include any reference to previous Macedonian emperors, even though he had most probably come across them in his sources. For instance, although *LT* describes how Basil I successfully conducted and supervised a river crossing with his army, the author of the *ST* did not include this anecdote, despite the fact that chapter 49.5 discusses exactly the same material and otherwise has many parallels with the relevant paragraphs found in *LT*.[36] This practice was by no means unprecedented; Nikephoros Ouranos, who is generally considered less creative with the treatment of his sources, did not include the name of Basil I in his treatise, even though he copied the respective passage from *LT*.[37] Given that there was most probably no such thing as the *Tactica Perdita* and that *LT* was most probably a direct source for the author of the *ST*, it seems rather odd for Constantine VII to have avoided such a reference to the founder of his dynasty, and even more so, when we bear in mind that Constantine played an important part in the compilation of the *Vita Basilii*, the laudatory biography of Basil I.[38]

These facts strongly imply that there was no dynastic connection between the emperor who commissioned the *ST* and Basil I, something which would exclude all the members of the Macedonian dynasty such as Constantine VII and Romanos II from being candidates for the authorship of the *ST*. Consequently, the strongest candidate to fit the profile of such an emperor is Romanos I. We have already discussed how the internal information of the *ST* fits well into the military and administrative milieu of his reign. But in addition to this, an attribution to him is the best way to explain the lack of reference to Basil I, which was actually more of a necessity. Romanos I was a usurper. Although he initially promised to give the throne back to Constantine VII when he was of age, he marginalized him for decades and finally gave precedence in the line of succession to his own son, Christopher. In this light, Romanos I would have been more than keen to avoid any dynastic memoranda linking back to Basil I, Leo VI and Constantine VII.[39] In addition, Romanos' commission may explain the quite independent style of the *ST*, which, as a whole, is not reminiscent of any other known manual. Now that the issue of authorship has been clarified, it is time to turn to the issue of attribution and attempt to determine why the title of the treatise features the name of Leo VI instead of Romanos I.

The attribution of the *Sylloge Tacticorum*

Undoubtedly, the false attribution of the *ST* has created much confusion and it is perhaps one of the main reasons why this manual has been viewed in a negative light. We could perhaps assume that the attribution is a later mistake or that it belongs to the long tradition of *pseudo*. The name of Leo VI in the title might have been an unsuccessful attempt of a scribe to identify the work which he was working on, the latter being perhaps copied from a manuscript that had a missing title. It could be that a scribe was confused by the similar material between the two treatises and thought he was indeed copying *LT*.[40] If the attribution was a mistake, however, it would suffice to have the name of Leo VI, as in the *Taktika*, not such a precise date along with it. It is very unfortunate that we do not have another group of manuscripts to make matters clearer; all we know for sure is that the *ST* in L was already attributed to Leo VI sometime in the fourteenth century. But now that the problem of authorship and sources has been discussed in detail, we can speculate about this attribution and put it in the context of tenth-century dynastic rivalry.

To begin with, it is almost certain that the title and attribution of the *ST* was a later addition. That does not only come from our previous discussion on the dating and authorship of the work but also from the very title itself. The title of the *ST* refers to Leo VI as 'αὐτοκράτωρ' and bears the date 6412 [903–904]. It has been noticed that before 904 the standard title that Leo VI and other emperors preferred and employed in documents was that of 'βασιλεύς'. It is only from an inscription dating in 904 and thereafter that we find Leo VI recorded as an αὐτοκράτωρ, probably in an attempt to distinguish himself from his brother and junior emperor Alexander.[41] This could therefore make the *ST* the earliest witness to this change, not to mention that 903–904 would make the compilation of the *ST* earlier than or contemporary with *LT*, something which is very improbable at best.

The fact that the title was added later and with such an unusual precise date seems to suggest that somebody deliberately wanted to associate the *ST* with *LT*. The *ST* seems to have been written in a time when the struggle for succession between Constantine VII, the legitimate heir of Leo VI, and the usurper Romanos I was still not concluded. Constantine VII was forced to suffer every humiliation in silence and therefore had every reason to hate Romanos I. The usurper had pushed him to the background for twenty-two years and expelled his mother and teacher from the palace. He had also allied with Nicolas Mystikos, the patriarch who had condemned Leo's wedding with his fourth wife, Zoe, the mother of Constantine.[42] Consequently it is very likely that Constantine VII or his circle changed the attribution of the manual to practise political propaganda. It is indeed a very possible explanation that the attribution of the *ST* is nothing but a *damnatio memoriae* of the name of Romanos I.[43]

Such an action is by no means unprecedented in the case of Constantine VII. Constantine had taken similar measures against his uncle, Alexander. When Alexander took the throne in 912 he wanted to make sure that he would be unopposed, and so he first attempted to eliminate Constantine as a potential threat by ordering his castration, something which in the end did not take place.[44] After Alexander's death, however, Constantine VII decided to impose a *damnatio memoriae* of the name of Alexander. In constitution 20 of *LT* an acrostic is formed which originally recorded the names of both Leo VI and Alexander. Constantine intervened in the text, and by making some slight changes in the first words of specific paragraphs he corrupted the acrostic only where the name of Alexander originally appeared.[45]

Similarly, Constantine VII also intervened in a funeral poem dedicated to Leo VI, in which he revised the part that originally referred to Alexander as 'ὁ τῆς πορφύρας ἥλιος'.[46]

If Constantine VII was so hostile to the memory of a member of his own dynasty, there was no reason to treat Romanos' memory any better. Indeed, the shift of Constantine's attitude towards Romanos is well reflected in his works. In the first half of the *De Thematibus*, which was probably written while Romanos I was still on the throne,[47] we read that 'Romanos the ruler, the good and valiant emperor, rendered it [Seleucia] into a *thema*, and added width, length and greatness to Roman authority'.[48] However, the favourable tone disappeared after Constantine VII became the sole emperor in 945. Consequently, in the *DAI* and the *DC* the treatment of Romanos I is all but flattering:

> The lord Romanos, the emperor, was a common, illiterate fellow, and not from among those who have been bred up in the palace, and have followed the Roman national customs from the beginning; nor was he of imperial and noble stock (. . .) he was too arrogant and despotic (. . .) but out of a temper arrogant and self-willed and untaught in virtue and refusing to follow what was right and good.[49]

> Note that his ceremonial for the Broumalia was changed and it reached the point of ceasing to exist in the reign of the ruler Romanos since, on the pretext of piety and thinking that it was not right for the Romans to observe the Broumalion (. . .) he ordered that these ceremonies cease. He did not bear in mind those great and famous emperors of the past (. . .) but whatever he thought right was deemed law and canon and righteousness and piety.[50]

Romanos I was not only criticized in the works of Constantine VII but also in that of others which seem to have been commissioned by men of his circle. The *Life of St. Basil the Younger* was written sometime in or after the sole reign of Constantine VII.[51] Two of the proposed patrons of this text were men who enjoyed much favour in the court of Constantine.[52] The first, Basil Lekapenos the Nothos, was an illegitimate son of Romanos I. He was probably castrated at an early age by Romanos I and therefore chose to cooperate with Constantine VII. He supported Constantine's accession to the throne and earned a much esteemed position in his court, holding the title of *parakoimomenos*. Together with John Tzimiskes, he was also entrusted with a Byzantine army to fight against the Arabs in 958 and was a great patron of the arts and commissioner of manuscripts.[53] The second were the Gongylioi brothers, who are described very favourably in the text and were said to be regularly visited by St. Basil the Younger himself. One of them, Constantine Gongylios, held the title of *patrikios* and enjoying the full trust of Constantine VII, he was put in charge of the expedition against the Arabs of Crete in 949.[54]

Whoever the commissioner of the *Life of St. Basil the Younger* was, he also had Romanos I described unfavourably. The latter is reported as 'an avaricious woman-izer and a corruptor of the citizens' daughters', despite the fact he only had one known bastard son, Basil Lekapenos, who was born after the death of his wife.[55]

The negative image of Romanos I in the text should not be underestimated, since there is good evidence to support that the *Life of St. Basil the Younger* did not aim at a narrow audience of followers or monks but at the broader public. Gregory, the author of the text, lived in the capital and had close relations with elite families in Constantinople. Furthermore, the story itself takes place in Constantinople, addresses famous people and provides universal messages of piety and apocalyptic visions which would have been appreciated by a wider audience. Finally, the large number of manuscripts which preserve the work seem to agree with the idea that the text was popular reading.[56]

The hostility, however, was not only confined to Romanos I but was also extended to his closest associates who were also, of course, rivals of Constantine VII and the Macedonian dynasty. John Kourkouas was unsurprisingly also affected by similar hostility towards his name. Being one of the most loyal supporters of Romanos I, he suppressed two plots against the throne and remained in his post as leader of the Byzantine army for almost the whole reign of Romanos I. John had every qualification to be a very strong political opponent: he was very well educated, came from a rich family and was a very successful commander who had acquired enormous fame from his successful campaigns against the Arabs. His contemporaries compared him to Trajan and Belisarius, while a work now lost was dedicated to him and to his military campaigns.[57]

John Kourkouas and his family seem to have been treated considerably harshly after his dismissal and the fall of Romanos I. As soon as Stephen and Constantine Lekapenos were exiled from the palace and Constantine VII became sole emperor, the properties of the family were plundered.[58] Other figures who served Romanos I in influential positions seem to have been treated more leniently. For example, Pantherios, who succeeded John Kourkouas in his post, does not seem to have received such harsh treatment, even though he was most probably a member of the Skleroi family who also cooperated with Romanos I.[59]

This discrepancy might be because the Kourkouai had a long-lasting rivalry with the Macedonian dynasty which dates back to the coup of John's grandfather against Basil I. The coup seems to have been very well organized and dangerous since sixty-six *archontes* and members of the senate also participated in it. However, this was not how Constantine VII wanted the events to be remembered. In the *Vita Basilii* he deliberately compressed and downgraded the coup, merely referring to it as 'one going by the name of Kourkouas, overcame by the lust of tyranny (. . .) gathered a band of like-minded plotters and waited for the opportune occasion'.[60]

In turn, the patron of the *Life of St. Basil the Younger* took the hostility to the next level. With reference to the Russian attack of 941 against Byzantium, the text records that those who defeated the Rus were Bardas Phokas, Pantherios and Spongarios.[61] This is, of course, contradicted by all our other surviving sources who state that the figures responsible for defeating the enemy were John Kourkouas, Theophanes and Bardas Phokas.[62] It is evident that the patron of the work applied a *damnatio memoriae* of the name of John Kourkouas and Theophanes, leaving intact only the name of Bardas Phokas, who came from a family hostile to the

Lekapenoi, who supported Constantine VII and who served as *domestikos* of the *scholai* during his sole reign.[63]

The same pattern is also observed for other close associates of Romanos I. For instance, Theophanes, a *patrikios* and *parakoimomenos* of Romanos I, who originally commanded the Byzantine fleet in 941 and who revolted against Constantine VII in a failed attempt to restore the Lekapenoi to the throne, saw his name replaced with that of Theodore Spongarios in the *Life of St. Basil the Younger*'s account of the Russian attack. Theodore Spongarios is either identified as the military saint Theodore or as a military official and *strategos* of the *thema* of Thrakesion.[64] The *Life of St. Basil the Younger* is also biased against Nicolas Mystikos, a close associate of Romanos I who was reluctant to approve the fourth marriage of Leo VI – the one which resulted in the birth of Constantine VII. Nicolas is recorded as having invited Constantine Doukas to share the crown with Constantine VII, but later as having betrayed him and as refusing him entrance to the palace, which eventually resulted in his death. Thus, Nicolas is described as having failed in his post and as worthy of condemnation by God for his actions.[65]

Under such circumstances, it is very probable that the attribution of the *ST* to Leo VI is, in fact, a *damnatio memoriae* aimed at Romanos I. Constantine VII was at first in a very weak position when he reached the throne in 945. Faced by rebellions, and given his military and administrative inexperience, he turned to the exercise of political propaganda through literature, in which he found a very effective way of highlighting his legitimacy and authority in matters of the state. This he did not merely by erasing and staining the names of his political opponents, as was the case for Alexander and Romanos I, but also by linking his name and that of his predecessors with the creation and preservation of handbooks related with the state and the army.[66] It is therefore most likely that Constantine VII and perhaps Basil Lekapenos changed the attribution of the *ST* to cover up the name of their political opponent in order to deprive him of such an authority, preferring to replace it with that of Leo VI. Both men were hostile to Romanos I and his associates. Moreover, they had both shown an interest in military treatises, new and old, as the manuscripts *Laurentianus Plut.* 55.4 and *Ambrosianus B.* 119 Sup. demonstrate, and they both practised political propaganda and *damnatio memoriae* through literary works. Constantine VII did so through the *DAI*, *DC*, the acrostic of *LT* and the funerary poem of Leo VI. Basil Lekapenos worked through the *Life of St. Basil the Younger* and perhaps also through the history of Theophanes Continuatus, as he has been credited with writing part of book VI and for its anti–Romanos I remarks.[67]

The dominance of the Macedonian dynasty in the years to come meant that this sleight of hand could go unchallenged. The Lekapenoi did not play a prominent role in Byzantine history after they were dethroned, and their only member who was active in the next decades, Basil Lekapenos, was a sworn enemy of Romanos' dynasty. Similarly, after the Kourkouai failed to dethrone Basil I, they seem to have never acquired any important office while a member of the Macedonian dynasty was on the throne. They were in disfavour until Romanos I came to power, and despite the fact that Theophanes Continuatus records that Constantine VII was willing to restore John Kourkouas' property and that he also entrusted him with the task of conducting one of the customary exchanges of prisoners between the Byzantines and the Arabs, it is no coincidence that the Kourkouai would only hold important posts again in the reign of Nikephoros II Phokas.[68]

The fact that Constantine VII and/or Basil Lekapenos seems to have been responsible for changing the attribution of the *ST* raises the question of whether they were also responsible for intervening in other parts of the text. This becomes more pressing when we take into account that the inconsistent cross-references of the *ST* strongly imply that the material of the manual was revised by some later redactor. It is therefore necessary to explore whether these two figures intervened with the manual and to estimate the degree and character of this intervention.

The redactors and the revision of the *Sylloge Tacticorum*

The relatively late date of L and the fact that only one recension survives present a huge obstacle in identifying the original or other versions of the *ST*. Consequently, it is very difficult, if not impossible, to prove for certain when or to what extent the manual was revised. Based on what is known for Constantine VII and Basil Lekapenos as redactors, however, we might be able to provide some speculations about the degree of their intervention in the *ST* by comparing the method they followed in other works.

As we have seen, the inaccurate cross-references in the *ST* point towards a later revision of the material because all the information does, in fact, exist in the text, but in the wrong order. It is impossible to reconstruct how the original material might have looked. The table of contents can be of little help here since it was added later and preserves the order of the chapters after the revision took place. In turn, the *pinax* implies that this revision occurred at some point between the compilation of the original version of the *ST* and before or at the time that somebody included the manual as the first work of a codex which also contained the *Poinalios Stratiotikos Nomos* and the *Akolouthia*. Judging from the inaccurate cross-references and where the correct material is indeed found in the text, we may speculate that originally chapter 35 was chapter 17, chapter 44 was chapter 20, chapter 23 must have been chapter 33 and probably chapter 38 was before the current chapter 22. This seems to tell us very little of how the original treatise might have looked like though, except for the fact that some of the material was somehow re-ordered and that the redactor did not return to correct these inconsistencies.

This careless re-ordering of material of an existing military treatise was something also probably practised by Constantine VII and Basil Lekapenos, who intervened in previous versions of *LT*. The version of *LT* found in the *Laurentianus Plut.* 55.4 has three constitutions extracted from the main body of the work, while the *Ambrosianus B* 119 Sup. has one constitution extracted. Although the remaining constitutions in both manuscripts were re-numbered, internal cross-references and the prologue of the treatise which refers to the contents of all twenty constitutions had not been revised to reflect these changes.[69] The constitutions of *LT*, extracted by Constantine VII and Basil Lekapenos, were given new independent headings. Apart from this, however, and some light changes to the text, especially in constitution 20, where the acrostic was corrupted, the redactors do not seem to have intervened greatly with the material of the work. Most of the treatise seems to be as Leo VI intended it.[70]

This method of re-ordering the material, of giving new chapter headings and of lightly intervening in the main text, may be helpful in recognizing similar patterns in the redaction of the *ST*. We have already seen that the redactors were most probably responsible for changing the attribution of the *ST* for political reasons, and in this light it is probable that they also made some light revisions to the text to the same end. This could mean that in chapter 53.8, where we read that siege towers should be countered with liquid fire shot through hand-siphons, the part that states 'the very thing which our Majesty presently invented' could have been a later revision, since *LT* 19.64 credits Leo VI for this invention. A comment such as this seems to be against the method of our author, who, as we have seen, seems to have directly copied something only in insignificant cases, in which the meaning or current state of affairs seems to be uninfluenced by such an imitation. Furthermore, Romanos I would have been keen to avoid any reference to the Macedonian dynasty and there is no other to be found in the text. On the contrary, this comment served the end of the redactors since it indirectly enhanced the attribution of the text to Leo VI.

As far as intervention in chapter headings is concerned, the re-ordering of the material must have created the need for fewer or more chapters and thus for new headings. We have already seen that some chapter headings in the *ST* demonstrate a characteristic which often appears in works that Constantine VII was involved with. Out of the 102 chapter headings of the manual, 35 begin with 'Ὅτι' and another 9 include it as part of the heading. What is more interesting, however, is that this phrase never appears in the main text, just in the chapter headings. The only exception to this is chapter 3.4, where we read that 'Ὅτι ἡ σχοῖνος ἑλληνικόν ἐστι μέτρον ταὐτὸ τῷ παρασάγγῃ'. It seems possible, therefore, that this was not the preferred style of the original author. It may well be that these parts which have 'ὅτι' in them, namely, chapter 3 or at least part of it, as well as the aforementioned headings, were a product of Constantine's or Basil's involvement, for Basil also preferred the use of ὅτι, as his version of the *Hypothesis* seems to demonstrate.[71]

Similarly, the fact that an independent title appears in the *ST* after chapter 75 could mean that this section could have had a separate existence and that it was later added to the *ST* by the redactors. This second title reads 'Στρατηγικαὶ παρανέσεις ἐκ πράξεων καὶ στρατηγημάτων παλαιῶν Ῥωμαίων τε καὶ Ἑλλήνων καὶ λοιπῶν ἐν κεφαλαίοις κή' and introduces us to the last part of the manual that preserves anecdotes from commanders of antiquity which originally derive from Polyaenus. The author seems to ignore this section in his introduction because while he states that he will first 'recall the armament and formations of former ages' and then he will 'place greater emphasis on those which are contemporary', he only partly does so.[72] Indeed, the treatment of Ancient Greek armament and deployment comes before the contemporary Byzantine one, but chapters 76–102, which are filled with ancient stratagems, appear after the contemporary material.

The fact that these stratagems are an abbreviation of Polyaenus could also mean, although not conclusively, that they were added by Constantine VII and Basil Lekapenos. In his treatise on imperial military expeditions, Constantine VII suggests that the book of Polyaenus was among the best to accompany the emperor during a campaign, while a Byzantine abbreviation of Polyaenus, the *Hypothesis*, was among the works included in the *Laurentianus Plut.* 55.4.[73] Basil Lekapenos also included a version of the *Hypothesis*, although different, in the *Ambrosianus B* 119 *Sup.*, better known as *Stratagemata Ambrosiana*. Their aim could have been to make the content of the *ST* more effective by providing the general with classical exemplars derived from an author they thought highly of. In fact, some of the exempla are very relevant to themes discussed in the first fifty-five chapters. For instance, chapter 100 refers to anecdotes regarding moderate punishments of soldiers, a topic which is discussed in chapter 17. Chapter 88 provides ancient examples of generals who personally acted in times of need, much as chapter 5 had already instructed the general to do so.

Apart from these, however, there seems to be another passage of the *ST* that does not belong to its author. In Dain's edition, after chapter 46 comes a *scholion* which is found in the beginning of the left lower margin of folio 94v and ends at the bottom margin of the same folio. The comment attempts to supplement the main material by drawing information from an ancient unspecified work which referred to the correct appointment of commanders for the right and left wings of the formation and explained which flank should take precedence during the crossing of defiles.[74] The fact that Dain included this comment in his edition seems to imply that he regarded it as an integral part of the treatise, but unfortunately he did not comment on its existence.

We cannot be sure when exactly the *scholion* was added to the text's tradition. It seems very unlikely to have belonged to the pen of the author though, since there is an obvious difference in style. In the comment, the commentator uses the phrase κίνησιν τῶν ἀνθρωπείων σωμάτων and the word divided in σώμασιν to denote marching and drawing up in units, respectively. However, throughout the manual our author uses the words τάγμα or τάξις to describe the various units; he never uses the word σῶμα in such a context.

A possible candidate for the authorship of the *scholion* are the copyists of L. All the other notes found in the margins of the *ST* are very short, though, save for this one, and those which comment on and fill the missing chapters 68–74. The vast majority of marginal notes act as markers, noting the subject treated in each paragraph of the main text, and none attempts to comment on the text supplementing it with knowledge from other works.[75] Consequently, the size, type and style of this *scholion* does not seem to fit the pattern of how the copyists of L interacted with the text. It could be that the comment originally belonged to some other scribe or owner of a previous manuscript which directly or indirectly served as a prototype for L, but it is also possible that it belonged to the redactors.

The phrasing and type of the *scholion* bring to mind another tenth-century example: the marginal comments which appear in the *DC*. Some scholars have argued that in these comments we are more likely to see a personal involvement of Constantine VII himself, especially in those which use the phrase 'χρή εἰδέναι'.[76] It may also have been, however, that Basil Lekapenos was somehow also responsible for them, since he most probably played an active part in the drafting and revision of the *DC*.[77] Whatever the case, the similar phrase 'δεῖ εἰδέναι' is found in the first sentence of the *scholion* of the *ST*, which is not only very close to 'χρή εἰδέναι' but also known to have been used interchangeably; for δεῖ usually replaced χρή when works were copied.[78] Furthermore, the commentator of the *ST* used the phrase Ἐκ τῶν παλαιῶν τακτικῶν to open his *scholion*, using information found in older tactical treatises to comment on the text. A similar practice, as well as phrasing, can be spotted in one of the comments of the *DC*, which opens with the similar phrase 'Ἐξ ἑτέρου παλαιοῦ τακτικοῦ' and also uses information from older treatises to comment on the main text.[79]

Consequently, it seems that the methods of revision undertaken by Constantine VII and Basil Lekapenos in other tenth-century works have some similarities with the redaction of the *ST*. The redactors most probably changed the original attribution of the *ST* and were responsible for light additions in the main text, as well as for the re-ordering of material and for inserting new titles. It is also possible that they added the last part of the *ST*, which preserves ancient stratagems, or it could be that these were originally scattered in the text and they decided to group them in a special section with its own title. Whatever the case, the redaction of the *ST* has the same unfinished character as other works which were revised by Constantine VII and Basil Lekapenos: its introduction and internal cross-references were not updated to correspond to the changes, and although the title of the last section states that the stratagems will cover twenty-eight chapters, one of them is missing.

Conclusion

The very precise attribution of the *ST* to Leo VI is mainly to blame for the variance of opinion regarding the value of the *ST* as a military manual and a source. While there is currently no consensus on who compiled the *ST*, the style, tone and phrasing of text itself seem to show that it was indeed commissioned by an emperor. The emperor responsible for the *ST* seems to have been a man who wanted to avoid any connection with the Macedonian dynasty. Contrary to texts commissioned by Leo VI and Constantine VII, he made absolutely no reference to emperors such as Basil I, even though such references existed in the material he used. The emperor most likely to match this profile is the usurper of the throne and enemy of the Macedonian dynasty, Romanos I.

In this context of political rivalry, the attribution of the *ST* to Leo VI appears to have been a *damnatio memoriae* of the name of Romanos I, probably imposed by Constantine VII and/or Basil Lekapenos who regularly practised political propaganda through literary texts. Constantine and/or Basil, however, seem to have further altered the text of *ST* in a way which agrees with their intervention in other military manuals. Material was moved around, new chapters and chapter headings were created, comments were added and the content of some passages was altered slightly. After this interference, the *ST* had the same unfinished character common to works edited by Constantine VII and Basil Lekapenos.

Notes

1 For an overview of previous views on authorship see the Introduction, pp. 5–8, and Dain 1938: 8; Dain and Foucault 1967: 350–1; Krentz and Wheeler 1994: xxi–xxii; Haldon 2014: 67–8. The issue of the sources and careless copying is discussed in Chapter 2.

2 The Greek reads τῆς βασιλείας ἡμων. The phrase is found several times in the manual: *ST,* 27.1, 35.2, 5, 36.2, 50.4, 6, 53.8 and 54.1. See Vári 1927: 266–7. Dain believed that the phrase 'our Majesty' could not be taken into consideration to clarify the issue of the authorship. Since the phrase appears only in the first half of the treatise, Dain envisaged it as belonging to the pen of the author of a non-extant source, the *Tactica Perdita*, and as being carelessly copied by the author of the *ST*. See Chapter 2 and pp. 5–8 above.

3 See for example: Leo, *Novels*, 40.71, 106.22; *LT*, 15.33, 2.21, 30, 4.1, 45, 11.9, 12.37; *DAI*, 45.68, 75; *DC*, 484, 528, 565; Daphnopates, 5.43, 4.60; LD, 115.

4 *TNO*, 61.16 (trans. McGeer, p. 119).

5 Δεῖ/χρὴ τον στρατηγόν.

6 Ταῦτα δὲ καὶ τα τοιαῦτα ἔτερα, (. . .) κρίνοντες δεῖν, τῷ παρόντι συγγράμματι συντετάχαμεν (. . .) ἵνα (. . .) οἱ ἡμέτεροι στρατηγοί, ταύτας φυλάττεσθαι ἔχοιεν (trans. Chatzelis and Harris, p. 94).

7 Vári 1927: 266–7. More examples of the same tone can be found in *ST*, 46.13, 47.20, 48.7.

8 See for example *MS*, 2.17, 2.20; *PM*, 1.1, 1.10, 1.94; Dagron and Mihăescu 1986: 164–5; Sullivan 2010a: 153.

9 (trans. Chatzelis and Harris, pp. 88–9).

10 Sullivan 2010a: 153; Dagron and Mihăescu 1986: 164–5.

11 See for example *ST*, 8.2 and c.f. *ST*, 20.1 with *MS*, 2.17.

12 Ταῦτα δὲ καὶ τα τοιαῦτα ἔτερα, οὐχ ὡς ἐνεργεῖσθαι παρ' ἡμῶν κατὰ τῶν πολεμίων δεῖν κρίνοντες, τῷ παρόντι συγγράμματι συντετάχαμεν (ἀνάξια γὰρ ἐμοί γε ταῦτα δοκεῖ χριστιανικῆς καταστάσεως καὶ μόνον λεγόμενα) (trans. Chatzelis and Harris, p. 94).

13 McGeer 1995a: 184–8; Haldon 2014: 67–8.

14 Vári 1927: 266; Dain 1938: 6–8; Dain and Foucault 1967: 357; Krentz and Wheeler 1994: xxi; Haldon 2014: 67.

15 Vári 1927: 268–70.

16 Irigoin 1959: 178–81; Dain and Foucault 1967: 382–5.

17 This subject is controversial and complicated. Different styles are noted in different works, and sometimes in different chapters of the same work. Some studies that treat this topic are: Moravcsik 1938: 514–20; Lemerle 1971: 276–7; Haldon 1990a: 70–5; Ševčenko 1992: 182–94; Tanner 1997: 128–30; Anagnostakis 1999: 97–123; Featherstone 2012: 123–35; Mango 2011: 3–13. Koutava-Delivoria 2002: 365–80, has argued that the contribution of Constantine VII in the *Geoponika* is perhaps underestimated.

18 *ST,* 1.24 (trans. Chatzelis and Harris, pp. 24–5).

19 Haldon 1990a: 178–9; Huxley 1980: 37–40.

20 Constantine VII, *Three Treatises,* C.8–10; *DC*, 510.20, 566.13.

21 Daphnopats, 61; Dujčev 1978: 237, 265, 281, 290–1, 294; Jenkins 1966: 297–8; Stavridou-Zafraka 1976: 368, 376, 389; Magdalino 1987: 58; 2013: 196–7; Dagron 1984: 268–9; Tougher 1994: 171–9; 1997: 126–8; Shepard 2003a: 341–4; Anagnostakis 2008: 45–60; Riedel 2018: 118–21.

22 Κατὰ Ῥωμαίων διάλέκτον / τῇ Ῥωμαίων διαλέκτῳ. See for example: Constantine VII, *De Thematibus*, 1.11.1–8, 1.6.1–4; *Geoponika*, 1.5.3; *DAI*, 29.216–17, 271–2; *DC*, 413.4–9; Koutava-Delivoria 1991: 281–332; 2002: 372.

23 *ST*, 44.4 and 45.11: Συνεκμπέμπειν δὲ ἄρα τηνκικάδε δεῖ καὶ τοὺς σκοποὺς μὲν καθ' Ἕλληνας, ῥωμαϊστὶ δὲ βίγλας ὀνομαζομένους. Τὸ μὲν οὖν βάθος ἤ πάχος ἐνταῦθα ἑκάστου τάγματος, ὅ δὴ καὶ κουντουβέρνιον ῥωμαϊστὶ λέγεται, ἐξ ὀρδίνων συνέστι δεκέξ (trans. Chatzelis and Harris, p. 61, 64). See also 40.7, 55.1, 55.4.

24 *LT*, 4.7 and 20.174: οὕς οἱ πρὸ ἡμῶν, νεώτεροι δὲ τῶν ἄλλων, τακτικοὶ 'Ρωμαϊστὶ καντάτωρες ἑκάλουν; διὰ τῶν καλουμένων τῇ ῥωμαίᾳ γλώσσῃ μινσωτατώρων καὶ ἀντικηνσώρων' (trans. Dennis, p. 51, 597).

25 *Geoponika*, 1.14.11 (trans. Dalby, p. 68); Koutava-Delivoria 2002: 368–9.

26 Constantine VII, *De Thematibus*, 1.17.1–4.

27 Kaldellis 2007: 13–188.

28 LD, 68 (trans. Talbot and Sullivan, p. 118).

29 For example: Constantine VII, *Three Treatises*, C420; *DAI*, 15.1, 21.3; *DC*, 520.12, 522.15; *Geoponika*, 14.7.28; Bury 1906: 538–9; 1907: 223–4; Haldon 1990a: 42–3; Sullivan 2003a: 145; Moffatt and Tall 2012: xxxii.

30 *ST*, 13, 93; Sullivan 2003a: 145.

31 Bury 1906: 538–9; 1907: 223–6, 428, 438; Haldon 1990a: 42–3; Németh 2010: 266–8; Moffatt and Tall 2012: xxxii.

32 *Mémorandum*, 1–4, 6–30, 32; Nicomachus, 1, 3, 4–5; *Suda*, π.323.13, 1941.17; Symeon Magister, 50.11, Leo, *Novels*, 7.8, 8.11, 10.38; Dain 1940a: 136; Dain and Foucault 1967: 366–7; Sullivan 2003a: 140–5.

33 Holmes 2010: 64–8; Magdalino 2013: 194–5, 201–9; Markopoulos 1994: 160–7.

34 *LT*, 9.14, 18.95; *DAI*, 30.128, 50.225; *DC*, 485.20; Constantine VII, *De thematibus*, 1.10.11, 11.21, 2.11.34.

35 See for instance: Constantine VII, *De Thematibus*, 1.9. *DAI*, 32.78, 40.8; 50.101 43.19, *DC*, 410–11, 514.19, 702.

36 See *ST* with *LT*, 9.12–18 and *MS*, 1.9.

37 Dain 1937: 46, 55.

38 For the debate on the degree of Constantine's involvement in the work see the studies of: Moravcsik 1938: 519–20; Lemerle 1971: 274–5; Toynbee 1973: 582–6; Huxley 1980: 30–1; Ševčenko 1992: 184–6; Anagnostakis 1999: 101–9; Kazhdan 2006: 137–44; Mango 2011: 3–13.

39 TC, 414; Skylitzes, 213, 216; Symeon Magister, 136.38, 50; Runciman 1929: 66–7; Toynbee 1973: 9–10. For numismatic evidence see Grierson 1973: 526–40.

40 Dain 1937: 53.

41 Spieser 1973: 162; Schminck 1986: 92–4; Haldon 2014: 59–60; c.f. van Bochove 1996: 44–5 for one of the exceptions in Leo's first novel, where both titles appear.

42 Symeon Magister, 136.9–13; TC, 397–8; Runciman 1929: 61, 65; Toynbee 1973: 9–14.

43 For some remarks on the Byzantine method of *damnatio memoriae* see: Vatchkova 2011: 164–6.

44 Symeon Magister, 133.4; *DAI*, 50.196–200; Karlin-Hayter 1969: 585–96; Ševčenko 1969–1970: 222–6; Grosdidier de Matons 1973: 241–2.

45 Grosdidier de Matons 1973: 232–40; Haldon 2014: 418–19; c.f. Schminck 1986: 97, n.271. For the religious significance of the acrostic see: Riedel 2018: 60–2.

46 Ševčenko 1969–1970: 202, 205–10; Grosdidier de Matons 1973: 241–2; Tsamakda 2002: 25–8.

47 Ostrogorsky 1953: 38; Huxley 1980: 31–2. For a presentation of the different arguments on dating see Lemerle 1971: 279–80.

48 Constantine VII, *De Thematibus*, 1.13, 2.6.

49 *DAI*, 13.149–57 (trans. Jenkins, pp. 73–5). See also Holmes 2010: 65, n.58.

50 *DC*, 606 (trans. Moffatt and Tall, p. 606)

51 Sullivan et al. 2014: 7–11.

52 Sullivan et al. 2014: 11; Magdalino 1999: 108–11; cf. Angelidi 2013: 25–6.
53 Ross 1958: 271–5; Brokkaar 1972: 199–217; Angelidi 2013: 11–26; Mazzucchi 1978: 267–31; c.f. Skylitzes, 288 who reports that it was Constantine VII who castrated Basil Lekapenos.
54 Skylitzes, 245–6; LD, 7.
55 *Basil Younger*, 1.29.15–21 (trans. Sullivan et al. p. 64); Grégoire and Orgels 1954: 153–4; Kazhdan 2006: 186–7 and Sullivan et al. 2014: 29 argue that the text is clearly biased against the Lekapenoi. Cf. Angelidi 1980: 170–1 who does not accept that there is such clear hostility. For Romanos and his morality see: Runciman 1929: 244.
56 Da Costa-Louillet 1954: 492–5; Angelidi 1980: στ, 86–91, 170–3; Kazhdan 2006: 186; Flusin 2001: 41–54; Sullivan et al. 2014: 19–24.
57 Symeon Magister, 136.10, 26, 76; TC, 397, 404, 426–7; Skylitzes, 211–12, 217, 230; Guilland 1950: 29–31; Andriollo 2012: 58–65; Howard-Johnston 1995: 87–8. For this lost work and the new trends in tenth-century Byzantine historiography see Markopoulos 2006: 397–405; 2009: 697–715.
58 TC, 441.
59 Cheynet 1986b: 146–7; Grégoire and Orgels 1954: 154.
60 Symeon Magister, 132.26; Leo Grammarian, 261; *VB*, 45 (trans. Ševčenko, p. 161); Guilland 1950: 30; Andriollo 2012: 58–9; Vlysidou 1985: 53–8.
61 *Basil Younger*, 3.23–27; Grégoire 1938: 292–3; Angelidi 1980: 161; Sullivan et al. 2014: 29–30
62 Symeon Magister, 136.71–5; TC, 423–5; Skylitzes, 229.
63 Grégoire 1938: 293–9; Grégoire and Orgels 1954: 154; Mango 1982: 306; Sullivan et al. 2014: 29–30; c.f. Angelidi 1980: 155–6, who argues that perhaps the name of Pantherios appears because he was in command the time St. Basil the Younger died, which coincides with the time that our author had perhaps started to draft his work. For St. Theodore, and his hagiographical tradition see: Haldon 2016: 1–19.
64 TC, 422–3, 430–1, 440; Skylitzes, 238–9. Grégoire 1938: 299; Angelidi 1980: 161; Grégoire and Orgels 1954: 153, argue that the role of Theophanes concluding the peace with the Hungarians is deliberately silenced in the *Life of St. Basil the Younger*. For the role of superstition in the context of Arab-Byzantine wars, see Ramadan 2014–2015: 1–37.
65 *Basil Younger*, 1.14–20. Gregory also shows a negative attitude towards the patriarch and son of Romanos I Theophylaktos, openly criticizing him (although Theophylaktos kept his post during the sole reign of Constantine VII). On the other hand, he is well disposed towards the Doukas family. He refers to Constantine Doukas as a martyr and states that he did not revolt, but came to the capital after being invited by Nikolas to share the crown with Porphyrogennetos. The Doukas family disappears after Nikolaos Doukas from the sources, which seems to indicate that it was pushed to the background and did not possess any high-ranking offices during the reign of Romanos. The only clear hostility at the time of Romanos I is reported from a certain Basil who revolted in 932, claiming to be Constantine Doukas, which shows that the family was still popular and a potential threat. The family reappears during the reign of Basil II. Grégoire 1938: 297; Grégoire and Orgels 1954: 148–50, 153–4; Da Costa-Louillet 1954: 495–6; Polemis 1968: 21–6; Angelidi 1980: 90, 117, 133–4, 137–9, 165; Rydén 1983: 572.
66 Holmes 2010: 64–8; Magdalino 2013: 194–5, 201–9.
67 Irigoin 1959: 178–81; Mazzucchi 1978: 267–316; Featherstone 2011: 115–23; 2012: 134; 2014: 353–72.
68 TC, 441–3; Vlysidou 1985: 56; Andriollo 2012: 66–75.
69 Haldon 2014: 55–66; Dennis 2014: ix–xi; Németh 2010: 30, 95–101; Tougher 1997: 168–9; Dain and Foucault 1967: 355–7; Mazzucchi 1978: 280–5.
70 Haldon 2014: 56; Grosdidier de Matons 1973: 229–31.
71 See for example: *ST*, 82.1; *Stratagemata*, 6.

72 *ST*, 1.1: ἀναγκαῖον οἶμαι [. . .] διαλαβεῖν τῶν τε κατὰ τοὺς ἄνω χρόνους καὶ ὁπλισμῶν ἐπιμνησθέντας καὶ παρατάξεων εἶτα μέντοι περὶ τῶν καθ' ἡμᾶς (trans. Chatzelis and Harris, p. 21).

73 Constantine VII, *Three Treatises*, C.196–199.

74 The comment reads: Ἐκ τῶν παλαιῶν τακτικῶν δεῖ εἰδέναι ὡς ἀεὶ τὸ ἀριστερὸν πλείονα τὴν ἐκλογήν ἔχει τοῦ δεξιοῦ. Διὰ τοῦτο καὶ τοῦ ἀριστεροῦ ἡγεμὼν διαφορώτερος ὀφείλει εἶναι τοῦ ἑτέρου ἐν στενοῖς δὲ καὶ δυσβάτοις τόποις ὁ τοῦ ἀριστεροῦ προηγεῖται, εἶτα ὁ τοῦ μέσου, καὶ τρίτος ὁ τοῦ δεξιοῦ. Προτιμᾶται δὲ διὰ τὰς ἐπελάσεις τοῦ δεξιοῦ τῶν πολεμίων τὰς κατ'αὐτοῦ. Εὐκινητότερα δὲ τὰ δεξιὰ τῶν ἀριστερῶν προηγοῦνται δὲ ἐν τοῖς στενοῖς τόποις κατὰ κίνησιν τῶν ἀνθρωπείων σωμάτων, ὅτε διὰ στενῆς εἰσόδου διελθεῖν αὐτά πρόκειται ἀνάγκη προτεθὲν δὲ ἐνταῦθα τὸ ἀριστερὸν μέρος πλάγιον διέρχεται πρὸ τοῦ δεξιοῦ μέρους, τὸ δὲ δεξιόν, ὡς εὐκινητότερον, καὶ συντόμως εἰς τὴν προτέραν ἀποκαθίσταται τάξιν, ὃ δὴ καὶ ἐν τάγμασιν ἐστιν ἰδεῖν παρακολουθοῦν.

75 See for instance the marginal notes on folios 81r, 82, 85v, 86, 89, 91v, 92, 93, 94.

76 Bury 1906: 538–9; 1907: 223–6, 428, 438; Haldon 1990a: 42–3; Moffatt and Tall 2012: xxxii.

77 Ševčenko 1992: 185–92; Featherstone 2004: 114–21; Brokkaar 1972: 218–19.

78 *DC*, 197, 686; Codoñer 2014b: 61–92.

79 Moffatt and Tall 2012: 161, n.3.

5   Tradition and originality in the *Sylloge Tacticorum* and its place in Byzantine warfare

Almost every literary work in Byzantium which was aimed at an educated reader imitated older models, both ancient and Byzantine. This, of course, generally applies to military manuals, and therefore to the *ST* as well, but did that *mimesis* simply reproduce word for word, or did it allow for innovations? This question concerns a wider debate about whether originality and innovation were present in Byzantine literature. To understand better tradition and originality in the *ST*, one must first appreciate the broader context and examine how our perception of Byzantine imitation has evolved over time.

Tradition and originality in Byzantine literature

It was a characteristic trend of older Byzantine scholarship to treat Byzantine literature as a genre without any innovation, a literature which stuck to ancient models and did not generally present the reality of its time.[1] Two characteristic examples are the views of Hunger and Cyril Mango. Hunger argued that 'the Byzantine Middle Ages cared very little for original genius' and with regard to military treatises and specialized literature that 'the Byzantines remained clung to ancient models'.[2] Mango stated that 'Byzantine literary works tend to be divorced from the realities of their own time, while remaining anchored in an ideal past' and that 'Byzantine literature is both a dim and a distorting mirror'.[3]

Since then, however, our understanding of innovation in Byzantium and the relationship between the empire and the past has been significantly improved. The current consensus is that innovation was present in almost every aspect of Byzantine life, like art and music, and that despite the fact that originality was not an end itself in Byzantine society, it was by no means absent from it.[4] The same can be said for literature – in spite of the evident *mimesis*, innovation can be spotted in many literary genres, such as historiography.[5]

This innovation can also be demonstrated in the study of inter-textual relationships between Byzantine literature and its sources. There it is proven that Byzantine *mimesis* was not slavish and that innovation can also be spotted in terms of how the Byzantines paraphrased their sources in order to fit things into their own context and to become original through the concept of *anakainises* and *ananeoses*, resulting in creating, as Ingela Nilsson puts it, 'the same story, but another'.[6]

It would suffice to provide some examples from tenth-century historiography and hagiography to support the case. To begin with, although the concept of φθόνος appears in the ancient sources of the *Vita Basilii* and the history of Leo the Deacon, both authors use the word with a Byzantine meaning of 'the devil/envious fate', which differs from that of their sources. Therefore, both Byzantine writers adapted the word to their needs and did not include it merely to imitate their predecessors.[7] As far as hagiography is concerned, although Niketas Magistros, the author of the *Vita* of St. Theoktiste of Lesbos, used the *Vita* of St. Mary of Egypt as a model, he largely revised and altered the narrative, creating new stories and manipulating the tradition.[8]

The kind of innovation that derives from revising and paraphrasing older models can also be spotted in the *ST*. As we have seen in Chapter 2, the author of the *ST* was creative with the treatment of his sources, and he usually adapted and paraphrased his models, regardless of whether these were classical or earlier Byzantine. We will attempt to study and identify these novelties, as well as determine the place that the *ST* holds in relation to other tenth-century treatises in terms of evolution. The innovations in the *ST* can be grouped into five groups: the adaptation of Onasander, the adaptation of Polyaenus and Julius Africanus, the presentation of updated attitudes to warfare, new tactics and battle formations and developments regarding technology and equipment.

Adapting and updating Onasander

One of the most notable adaptations undertaken by the author of the *ST* was to insert a number of comments and passages which are in accordance with a distinctly Christian character.[9] These passages appear as an addition to Onasander and serve to make his arguments more relevant to a Byzantine context.[10] Most of these passages concern certain traits and actions that the ideal Byzantine general should have or use.

One such example is found where the general is advised to be careful in times of truce and to avoid dropping his guard against the enemy. Onasander had originally argued that the general should 'suspect a breach of faith on the part of the enemy due to their hostility'.[11] In chapter 16.3, however, the author of the *ST* revised this statement, stating that 'from the side of the enemy dishonesty should be suspected because of barbarian morality'.[12] Consequently, distrust is connected with moral issues, which most probably derive from the fact that the main enemy of the Byzantines, the Arabs, were Muslims. The theme that the Muslims were not be trusted for long also appears in the contemporary imperial correspondence, written by Theodore Daphnopates on behalf of Romanos I.

Daphnopates was in a very esteemed position in the court of Romanos, serving as *protasekretes* and also bearing the title of *patrikios*.[13] In his letter, which was addressed to Symeon I and was probably composed around 925–927, Daphnopates tries to convince the Christian khan of the Bulgarians to make peace with Romanos, writing 'since you are a lover of the perfect and true religion, why not agree to peaceful and perpetual terms?' Daphnopates urges Symeon 'not to appear worse than barbarians [Arabs] in our disposition, who even though they make truce and exchange prisoners, they do not accept complete peace since they lack faith in the perfect religion'.[14] The concept that the Arabs could not make a long-lasting peace with the Byzantines due to religious issues is also brought up by Arab sources. Around 957–958 some Byzantine emissaries visited the court of the Fatimid caliph al-Muʿizz requesting a perpetual truce. Qadi al-Nuʿman, who held a very prestigious position in the court of al-Muʿizz, reports that the Fatimid ruler responded that a perpetual truce is not in line with the Muslim religion and canon law, which preach holy war against the infidels and only allow a truce for a fixed amount of time.[15]

The addition of distinctly Christian elements to complement Onasander's thoughts can also be spotted in chapter 1.7, which is, in fact, declarative of how paradoxical and interesting the relationship of Byzantium with the classical past was. In this passage the author of the *ST* discusses the traits of the ideal general, one of which is prudence. At first, he copies Onasander almost verbatim, stating that the general should be 'prudent, so that he might not abandon his devotion to the most important [things], by being drawn to physical pleasures'.[16] Our author, however, supplemented Onasander with a passage from Basil I's *Hortatory Chapters* which states that God grants victory to those who are able to resist the pleasures of life, explaining that these men first win over their moral/invisible enemies, and therefore they also prevail against the visible ones.[17] While this addition served to highlight the typical Byzantine concept which connected Christian morality and piety with the ability to win in battle,[18] it is nonetheless substantiated not by some reference to Christian fathers, but with a rather suitable saying of a classical Greek philosopher, Aristippus of Cyrene, who wrote that

> He who prevails over pleasure, is not the one who refrains from it, but he who is not carried away by it despite the fact that he desires it. Just as, for instance, with the ship and the horse: [it is] not the man who does not desire them, but he who wants to travel somewhere.[19]

Last but not least, additions of Christian morality were by no means only made to complement the material of Onasander, but some were included to accompany novel characteristics which were added anew to the portrait of the perfect Byzantine general. It comes as no surprise that two of these traits were piety and justice.[20] The author of the *ST* supplements the latter with comments on the Last Judgement stating that 'whoever governs lawfully, but is tolerant of those who act unjustly will be judged by God with the same measure as those who are unjust'.[21] Apart from religion, however, the addition of justice can also be linked with the administrative duties of the *strategos* in the tenth century. The general was responsible for military rewards and punishments, but the *strategos* also had supreme jurisdictional authority over his *thema* and was active in judging cases until the mid-tenth century.[22] Moreover, the fair judgement on the part of the *strategos* was even more essential in a tenth-century context because of the need to enforce the novels of Leo VI, Romanos I and Constantine VII to protect the military lands of the *stratiotai* against the *dynatoi*. The author of the *ST* makes specific reference to this problem twice, slightly adapting the text of *LT* which was nevertheless still relevant. More specifically, the *ST* states that the general should particularly protect farming and military matters and also stresses that the *stratiotai* should be free of impositions and extra payments.[23]

Adapting and updating Polyaenus and Julius Africanus

Except for adapting Onasander, the author of the *ST* or a lost source he could have used has further manipulated the ancient authorities. The *ST* features various cases in which some aspects of classical stratagems are changed.[24] Some of these changes seem to have been undertaken with the aim to make stratagems more relevant to a Byzantine context.

For instance, the author of the *ST* copied the advice of Julius Africanus in chapter 59, describing how to overcome the enemy through poisonous food. Although the version of the *ST* does not add anything new to the tradition, a certain originality is identified in chapter 59.3 where such practices are labelled as 'unworthy even to be said in a Christian context'.[25] The author of the *ST* continues to explain that this information was included not to be used by the Byzantines, but simply to keep the general informed of the practices that may be used by the enemy. It could be that the author of the *ST* reflected contemporary attitudes. The use of poisonous food and water against the enemy features as a standard practice in *MS*, but in *LT* such action is no longer advisable to be undertaken by the Byzantines. *LT*, copying *MS*, only states that the general should guard against such practices and suggests that the Byzantines used poisonous arrows mainly, if not merely, against horses and not riders.[26] Historical narratives seem to support this approach since no relevant evidence is found.

Second, some stratagems became more relevant through the addition of a number of anachronistic references to Byzantine technology. In chapter 65, although the *ST* reproduces Julian Africanus' original advice to fall upon the enemy cavalry with burning torches in order to make it flee, it goes further to add that this can also be achieved by spraying the horses' nostrils with spurge juice through a hand-siphon. This machine was only invented in the reign of Leo VI, centuries after Julian's death.[27] Similarly, in chapter 99.5 we find reference of added protection for the horse's hoofs, a technological invention which was probably in use by the Byzantines around the ninth century.[28] The latter is not only important because of the anachronistic reference, but also because it appears in a brand-new stratagem which seems to be absent from the tradition of Polyaenus.

It appears, in fact, that a tradition was created out of thin air with a view to presenting the instructions of the treatise as belonging to the classical past.[29] The *ST* features new figures and stratagems, some of which have a more or less realistic historical background, like that of Choerillus, while others are at best obscure and previously unheard of, such as Merops, Onias, Tyrrenius and Abradatas.[30] Deceptive as this fabrication of tradition may seem, it was by no means something unprecedented in Byzantine literature. Examples from religious texts demonstrate similar behaviour with a view to strengthening the authority of their material.[31]

Last but not least, there were cases where such drastic measures were unnecessary, since some stratagems became more suitable with almost imperceptible manipulation. A characteristic example is found in chapter 86.5 where we read of the stratagem of Brennus. According to the *ST*, Brennus presented the shortest and weakest Roman prisoners to his army in order for it to gain courage for the upcoming battle with the Romans. In the *Hypothesis*, the Byzantine epitome of Polyaenus, Brennus is recorded as having done the same thing, but we read nothing about the identity of the prisoners, nor about the enemy's. Going further back to the original text of Polyaenus, the prisoners are Greek rather than Roman and Brennus is a Gallic king fighting against the Greeks. Consequently, in the *ST* the tradition of Brennus has been manipulated from Brennus who invaded Macedonia and Greece in 279 BC to Brennus who attacked Rome in 390 AD. This served to make the stratagem more relevant to the Roman past and closer to the Byzantine experience, since the latter considered themselves the only true continuators of the Roman Empire.[32]

Updated attitudes to warfare

The *ST* was compiled at a time when the Byzantines started to get the upper hand against the Arabs. The tenth century was the time that the big families of Asia Minor came to prominence and started to enrich themselves through land acquisition and booty from the frontier wars. It was also the century that promotional literature of prominent generals started to appear, like that of John Kourkouas Bardas Skleros and Nikephoros II Phokas.[33] This type of literature presented its subjects in an ideal way, sometimes playing it by the book and practicing the well-known Byzantine indirect approach to war, but sometimes acting like warriors who considered pitched battles and duels honourable, who highly valued hand-to-hand fighting and bravery, and who saw wounds as a medal of honour and retreat as disgraceful, unless it was conducted in the most desperate situation. Although these works are now lost, their traces can be found in extant historiography from Theophanes Continuatus and the *Chronicle* of Symeon Logothete onwards. Therefore, the readers of such works, and especially the military aristocracy, were exposed to such heroic ideals along with the more prudent Byzantine tradition of indirect warfare.

These heroic ideals were neither new nor specifically Byzantine. When the conditions were ripe for them to become popular, they were expressed in accordance with the authoritative classical past. Authors once more turned to classical works which had long provided the Byzantines with this dual perception of warfare: one could either be a Ulysses, using cunning and stratagems to overcome the enemy, or an Achilles, fighting in the open with bravery and prowess in arms.[34]

The impact of these ideals was not confined to historiography to entrench and idealize. On the contrary, they seem to have influenced the military aristocracy more deeply, since they also appear in some military manuals. The *ST* is, in fact, the oldest among the manuals to preserve some of them so clearly. The first heroic ideal to concern us is bravery and death in battle, which was a very popular aspect of promotional historiography. For example, drawing on promotional sources from the tenth century, John Skylitzes records in the late eleventh that in 979 Bardas Phokas 'was of the opinion that it was better to die gloriously than to live ignobly'.[35] It is in *LT* that we first read that the general should be 'bold in the face of dangers' and that 'it is more beneficial to take a stand in battle (. . .) than to flee', but the *ST* further advances this case.[36] In the latter, bravery appears as one of the traits of the ideal general, and it constitutes an addition to the characteristics originally found in Onasander. What is more, originally drawing on Isocrates, the author of the *ST* continues to explain that bravery is a much esteemed trait, 'for death is common to all mankind, but to die gloriously is a characteristic of the great'.[37] Katakalon Kekaumenos makes a very similar case, urging in his manual not to 'fear death, if you are going to meet it on behalf of your country and the emperor; rather fear a shameful and blameworthy life'.[38]

The second heroic ideal was fighting openly, which was sometimes considered more honourable to fighting with stratagems. Promotional historiography usually underlined this mentality by giving examples of contradicting actions. For instance, Nikephoros Bryennios highlights how his grandfather fought openly and honourably in pitched battles, while in comparison Alexios Komnenos emerges as an ignoble general trying 'to steal victory' by fighting with ruses.[39] In the late tenth century, John Skylitzes reports that at first both Bardas Phokas and Bardas Skleros 'hesitated, and shirked open battle and attempted to steal victory', but describes their duel very honourably and full of praise.[40] Theophanes Continuatus records that Theophobos advised emperor Theophilos (829–842) to make a night attack, but this did not take place, as the emperor was convinced by the majority that Theophobos was trying to deprive him of his glory and they urged him to fight at dawn instead.[41] The latter brings to mind how Alexander the Great reacted before the battle of Gaugamela, in 331 BC, when Parmenion advised him to launch a night attack against the Persians, responding that 'it was dishonourable to steal the victory'.[42]

The *ST* is the first and only manual which preserves this ideal so evidently. In chapter 48.7 we read that 'it must be known that night battles were invented for times of weakness or shortage in the army. For if the army is fighting-fit, it [is] insulting and totally unworthy to win in such a way'.[43] It seems as if we are presented with the credentials a night attack should fulfil to be considered honourable. While this remark is not totally alien, since manuals generally advised the general to engage in pitched battles if his army was well prepared and stronger than the enemy's, there is, however, some contradiction with the advice found in other military manuals such as *LT* and *PM*, and it is questionable whether generals really took it into consideration when they conducted night attacks.[44]

Tactics and battle formations

Although we have spotted a number of innovations in the *ST*, so far these are literary and theoretical. They were included as an addition to or as a manipulation of the classical and Byzantine tradition in order to update it to a more contemporary context. But despite their literary side, military manuals also had their practical side. They seem to have been connected to a specific time which had certain needs to tackle and evolutions to respond to. They seem to have been created, at least to some extent, to be used by officers and generals, to provide them with guidelines on how to best use their limited resources, technology, ruse and geography to their favour. It is therefore necessary to look for another kind of innovation, a more practical one, and to search for originality and evolution in the realm of tactics and battle formations.[45]

As we have seen, the *ST* was compiled in a very specific context in which the traditional guerrilla practices were still predominant but started to co-exist with the new needs and challenges required by the Byzantine counter-attack and by the new role the army gradually started to play.[46] Despite these evolutions, guerrilla tactics remained a crucial aspect of Byzantine warfare at least until the reign of Nikephoros II Phokas.[47] With a view to providing the reader with an efficient guidance on these matters, the author of the *ST* included all the relevant information he could find on the topic from Onasander, *MS*, the *PS* and *LT*.

But although the *ST* is based on these authorities, we can recognize some special emphasis given on certain key aspects. For instance, in chapter 49.10–2, despite the parallel information found in *MS* and *LT*, our author provides us with two marching formations, designed to ensure a safe crossing of medium and narrow defiles.[48] Furthermore, in chapter 23.4, the problem of withdrawing after a raid is examined in detail, and the regular practice of the enemy to capture the *kleisourai* and to attack while the army was withdrawing is also emphasized. Although the author of the *ST* copied earlier information on how to defend against an invading enemy when there was a shortage of troops, which was a very important aspect during the yearly Arab raids, he further instructed the reader to attack while the enemy was setting up camp or while he was occupied with the pasturage of horses. He also warned that the latter stratagem is also undertaken by the enemy.[49]

Nonetheless, one can better recognize the contribution of the *ST* to Byzantine military theory and the gradual evolution of the latter by studying the appearance of new units and tactics. It is no surprise that these first appear in the *ST* since offensive warfare made the use of specialized troops essential.[50] These new conditions were what shaped tenth-century infantry and cavalry tactics as we know them, and they all began with the *ST*.

We have already seen that the *ST* is the first manual to list the medium infantry as a distinct type of infantry. This is a gradual evolution from *LT* where the javelin men were considered and operated as light infantry.[51] The new type of medium infantry is called *peltastai* in the *ST*. The term used is obviously antiquarian, and it automatically brings to mind the role of the classical *peltastai* from the Hellenistic period. The latter were a medium type of infantry, less heavily armed than the traditional hoplites, but at the same time more heavily armed than the light infantry, able to employ both melee and missile roles on the battlefield, as they were equipped with swords, lances and javelins.[52] But was this just a theoretical anachronism, or did it actually correspond to the needs of contemporary Byzantine warfare of the time?[53]

It appears that further specialization was in line with contemporary Byzantine needs, and this is corroborated by the testimony of manuals produced throughout the tenth century. In the *ST* the *peltastai* are described as bearing smaller shields than the heavy infantry and being equipped with a *paramerion*, javelins, a spear, a helmet which did not cover the face and either a *lorikion*, a *klibanion* or a *kabadion*. They appear to be somewhat lighter than the heavy infantry, as they had smaller shields and did not bear arm and leg guards.[54] This type of medium infantry, who qualified as a sort of heavier javelin men, is found in the later manuals, though not with the antiquarian term *peltastai*, but rather referred to as *akontistai*. This separation of infantry in three categories, rather than two, also appears in the *DV*, in the *DRM* and in the *PM*, all of which treat the *peltastai/akontistai* as a distinct type of medium infantry that was usually drawn up together with the heavy or supported the light.[55]

Another new unit, first found in the *ST*, is the corps of the *menavlatoi*. They were specialized infantry whose duty was to repel the attacks of the enemy heavy cavalry.[56] The *menavlatoi* were named after their weapon: the *menavlion*. Although the *menavlatoi* are first attested to in the *ST* as a distinct unit, the *menavlion/menavlon* as a weapon appears already in *LT*.[57] There is some scholarly debate concerning the length and the use of the *menavlion* which mainly results from the fact that before the *ST*, the word *menavlion* was used to describe a javelin and not a sturdy thrusting spear.[58] We can note that the *ST* appears to be the intermediate stop in the evolution of tactics, since it is the first manual to preserve the innovation of the *menavlatoi* and the change in the meaning of the *menavlion*, but the tactics and the numbers of the *menavlatoi* become further advanced and sophisticated in the *PM*.[59]

Nevertheless, the infantry was not the only one that was updated and specialized in the *ST*. As specialized infantry units like the *menavlatoi* appeared and the new role of the medium infantry emerged, similar evolution was also required in the cavalry. The *ST* is the first military manual to report the Byzantine *kataphraktoi*. Despite the fact that the Roman army started to employ units called *cataphractarii/clibanarii* already in the reign of Hadrian (117–138), they are last recorded in the sixth century, when there is a reference to a unit called *Leonis Clibanariis* in Egypt in 546.[60] The next available evidence of the term *kataphraktoi* to denote a specialized troop of heavily cavalry and not heavy armour or more heavily armed troops in general occurs in the *ST*, some 400 years later.[61]

The disappearance of the term *kataphraktoi* from the sources for about four centuries has aroused much debate concerning the tactics and the use of heavy cavalry in the Byzantine army. From the sixth to the early tenth century the Byzantine cavalry went through great changes, and warfare was dominated by the influence of the steppe nomads, the fluidity of the battlefield and the de-specialization in equipment.[62] The sources of this period – manuals, poems and historical narratives – feature a de-specialized cavalry mainly operating as horse archers or as archer-lancer cavalry.[63] The elite cavalrymen, together with their mounts, remained armoured, but the term *kataphraktoi* was not used to describe them.[64]

In light of this evidence, two different perspectives have emerged. Some argue that the *kataphraktoi* were disbanded from the Byzantine army since the defensive warfare of the period did not facilitate the use of such troops, only to appear again around the middle of the tenth century.[65] Others, on the other hand, argue that the *kataphraktoi* were never disbanded. Instead, they suggest that the appearance of the term *kataphraktoi* could be incidental, while they also underline that there are certain factors, such as authorship, trends and genre, which influence the information we receive in the sources. From this point of view, the revival of the term *kataphraktoi* could be connected with the literary style of the Macedonian renaissance.[66]

A more probable solution to the problem could be as follows.[67] The Byzantines never stopped employing heavy cavalry – there is, after all, mention of them in historical narratives and manuals. The fact that this cavalry was not specialized, however, resulted in the abandonment of the term *kataphraktos*. Wherever the term appears before the tenth century, it merely denotes armour or a heavily armed soldier in general, either infantry or cavalry. It is only from the *ST* onwards that, in addition to the general meaning, the term *kataphraktos* denotes a specialized heavy cavalry which employed distinct and specific tactics.[68] The emergence of the term *kataphraktos* to denote a specialist heavy cavalry, whose role was to employ shock tactics by directly charging against enemy formations, is in accordance with the new specialized infantry units that appear in the *ST* and with the operational needs of the tenth century.[69]

In the *ST* the *kataphraktoi* are presented in their very first stage of evolution. Judging from relevant information found in *LT*, it seems that their revival as specialized heavy cavalry troops took place gradually. The *ST* happens to be the first manual to preserve the wedge formation of the *kataphraktoi*. Although the wedge formation was known from antiquity, its composition of *kataphraktoi*, lancers and horse archers, as well as the pattern of adding four men in every rank, is a contemporary Byzantine invention first found in the *ST*.[70] The wedge formation of the *kataphraktoi* was placed in the middle of the vanguard to spearhead the attack, but that is not entirely unprecedented. In *LT* we read that 500 elite cavalry were to be placed in the middle of the vanguard. These 500 elite horsemen could have served as an inspiration for the wedge of the *ST*, for the latter did not only deploy the *kataphraktoi* in the same position but also one of its variations of the wedge formation numbers: 504 men.[71]

The appearance of a number of specialized new units also required a fresh approach to deployment and tactics which would allow the Byzantines to make their use effective. As far as the infantry is concerned, the author of the *ST* seems to have been aware of such a need, since he argued that a discussion about the deployment of infantry units was essential since the latter usually fought irregularly.[72] It could be, therefore, that the guerrilla tactics of the soldiers of the *themata*, which dominated Byzantine warfare, made the infantry unable to live up to the task of effectively fulfilling the new required roles. To remedy this, the author of the *ST* provided his readers with a number of new infantry formations.

The most important of these new formations is the appearance of a new standard battle array: the hollow-square formation. The author of the *ST* begins by making a case about its uniqueness, stating that this formation is 'undoubtedly dissimilar to the fashion of the aforementioned infantry and cavalry formations'.[73] This formation, however, had some known parallels from antiquity, and it also appears in earlier Byzantine manuals such as *MS* and *LT*, but only as a marching or emergency formation, never as a standard one, as is the case in the *ST*.[74]

In contrast to the brief treatment of this array in past manuals, the author of the *ST* describes it in great detail. The infantry was drawn up in a square, each unit leaving intervals between one another. The number of intervals depended on certain factors, such as the number of the cavalry or the numbers of the enemy. Inside the square the cavalry was drawn up together with the baggage train. The cavalry assumed the offensive role, exiting through the gaps to attack the enemy, while the infantry served as a mobile operation centre, providing defence, support and a rallying point if the cavalry was thrown back.[75]

After Leo VI had explained that a hollow-square formation can be used if the army is defeated and threatened by a cavalry army, he stated that the Arabs employ battle formations which are

> both square and oblong and so are very secure and not easily broken up by the attacks of their opponents. They employ this formation while marching and in forming up for battle. They also imitate the Romans in many respects. It is as though they have been trained by experience in the other models of battle formations, so the very things they suffered from the Romans they are now busily putting into practise against them.[76]

From this statement Leo VI seems to suggest that that the Arabs were already using the square formation in battle, as well as others, which they took from the Byzantines. In this light, we may argue that Leo VI presents a point in which the Arabs were partially using the square formation in battle, but the Byzantines had not yet started to employ it as their standard battle order, something which happens in the *ST*.[77] The hollow square in the latter seems to be in its very first stage of evolution as a standard battle order, for not only does the author of the *ST* describe it in detail, but its deployment does not seem to be crystallized. The *ST* is the only manual that deploys all the units of the square by ranks, which makes the flanks look rather vulnerable. This and other such issues were improved on in later manuals since the *PM* draws up the sides of the square by file rather than by rank and introduces additional beneficial changes.[78]

Apart from the hollow square, the *ST* describes another formation which was exclusively composed of infantry. We find four variations of this infantry formation ranging from 24,100 men to 3,116.[79] Although this formation is new, its core seems to have been evolved from an infantry deployment described in *LT*.[80] While there seems to be an evolution from *LT* to the *ST*, there is no further mention of such a formation in later manuals, and its practicality is questionable.[81]

In the *ST*, however, the new tactics and formations are not confined to the infantry, for similar developments are noticeable in the cavalry. The *ST* is the first manual to preserve a new cavalry formation. This is presented in various versions according to the available number of troops, but the most serviceable versions deploy the troops in three main lines. Once again, this innovation seems to have been based on a gradual evolution of what is found in *LT*, as the basic plan remained more or less the same, although with certain additions.

To begin with, both *LT* and the *ST* draw up the *prokoursatores*, together with one *tagma* of *defensores*, as scouts ahead of the main force and some small units as a separate rear guard.[82] As is the case in *LT*, the *ST* instructs the general to divide his vanguard into three main units, the only difference being, as seen earlier, that the middle unit in the *ST* took the form of the wedge of the *kataphraktoi*. Both manuals also agree that flank guards and out-flankers should be posted at the flanks of the vanguard as well as another two concealed units, one in each flank, which were to be posted farther away from the flank guards and out-flankers. With regard to the second line, both manuals suggest that it should consist of four main *tagmata* which were to keep larger intervals than the units in the vanguard. The larger intervals were to be guarded by other smaller *tagmata*. However, the author of the *ST* adds a new third battle line which is not found in *LT*. He goes on to explain that this line should be identical to the vanguard, if there are enough *kataphraktoi*; otherwise, its middle unit should consist of regular cavalry instead. This new third line is named *saka* by our author, which is the word that Muslim authors use to describe the rear guard. It has been argued that most probably this implies a Byzantine imitation and adaptation of tactics to contemporary Arab developments.[83] Whatever the case, once more, the *ST* acts as a middle point of development since later tenth-century manuals, such as the *PM*, adopt this cavalry formation with very few alternations.[84]

Technology and equipment

The *ST* does not only provide original material regarding tactics and formations, it also features some innovation and evolution in terms of technology and equipment. The character of this innovation is the same as was noted earlier. It is a gradual evolution in which the *ST* appears as the middle point of development between *LT* and other later manuals such as the *PM*.

The first novelty concerns siege warfare and the appearance of the *strepta*. It is open to debate whether the *strepta* was a different device from the hand-siphons, whether it was the same device or whether it was a different part of the same device.[85] Whatever the case, the *ST* seems to treat the *strepta* as a different device from the hand-siphons. Although the *strepta* also appears in the siege manual of Heron of Byzantium and in the *De Obsidione Toleranda*, accepting the new dating of the *ST* would mean that the latter is the first manual to mention this device.[86]

This seems to be supported by the fact that the author of the *ST* felt the need to describe the function of the *strepta* to his readers. He explains that the *strepta* is the device 'which mechanically shoots the liquid fire'.[87] Even if the *strepta* was identical to the hand-siphons, the *ST* is still the first manual to encourage the use of such devices for sieges, since in *LT* the hand-siphons are only referred to in the context of naval warfare.[88]

Except for the technology of siege warfare, similar innovations can be identified with regard to the equipment of the Byzantine infantry. On one level the equipment of heavy infantry in *LT* and in the *ST* has many similarities. For instance, the men are instructed to wear *lorikia*, arm and leg guards; to have plumes at their helmets and shoulders; and to carry circular shields, double-edged swords and lances. On the other hand, the *ST* recommends square and triangular shields in addition to circular ones. Not only are triangular and square shields treated as a standard type for the first time, but also detailed measurements are given for all. The axes that Leo VI instructed the heavy infantry to have are not found in the *ST*, which features *parameria* instead. What is more, the *ST* presents the general with a number of alternatives regarding the heavy infantry armour. In addition to the *lorikia*, the use of *klibania* and *kabadia* is recommended. The latter were made of cotton and raw silk, and the *ST* is the first manual to record their existence. Finally, we can also note that although full-face helmets are found in *LT*, Leo VI does not specify their use for the heavy infantry, whereas the author of the *ST* does.[89] As far as the armament of the light infantry is concerned, it does not present any innovations compared to *LT*, apart from the fact that the *ST* provides a precise measurement of the size of slings.[90]

A similar gradual innovation can also be spotted in the equipment of the heavy cavalry. Once more, the *ST* appears as the middle point of evolution between *LT* and the *PM*. The heavy cavalry in the *ST* is very similarly armed to the elite cavalry in *LT*; both manuals instruct that the heavy cavalry should be equipped with bows, *klibania* or ankle-long coats of mail, lances and shields. In the *ST*, however, the *kataphraktoi* are clearly instructed to bear fully covering helmets and also surcoats, called *epilorikia* or *epanoklibana*. In addition, while in *LT* the use of the mace was confined to a small portion of the infantry or to be carried by the latter as a secondary weapon in their waggons, in the *ST* it is first attested as a weapon of the heavy cavalry, though only as a secondary one. The equipment of heavy cavalry appears more crystallized in the *PM* which leaves out the bows, the long mail coats and the lances. Instead, the *kataphraktoi* are armed with fully covering helmets, *klibania*, *epilorikia*, shields and maces.[91]

Conclusion

The *ST* is a manual where innovation can be seen in many aspects. Except for original material which supplements and updates classical sources and provides new attitudes to warfare, there is the introduction of new troops, both infantry and cavalry, as well as evolution in their formations, tactics and equipment. These innovations paint the picture of a Byzantine army which was gradually evolved from *LT* and which was in the middle point of evolution between the *LT* and the *PM*. The appearance of specialized troops such as the *menavlatoi* or the *kataphraktoi*, the hollow-square and wedge formation and the use of new terms and equipment such as the *kabadion* or the *epilorikion*, as well as the broader use of the mace, were all important aspects of tenth-century warfare first recorded in the *ST*.

On these grounds it would be fair to claim that the *ST* is one of the most innovative military manuals the Byzantine world produced. Its innovations paint the picture of an army which had some similarities with *LT* but at the same time was totally different. It was an army which could fulfil its traditional guerrilla roles, but also an army which gradually evolved and was ready to correspond to the current offensive needs in the oriental front between the reigns of Romanos I Lekapenos and Nikephoros II Phokas. These tactics were further evolved in the *PM*, but it is reasonable to state that the *ST* is the manual that introduces us to tenth-century Byzantine warfare as we know it. After the *ST*, the Byzantine tactics were just not the same again; they had evolved into something new.

Notes

1 Reinsch 2010b: 56–7; 2010: 23–6; Nilsson 2010: 195–8; Kazhdan 1995: 8–9.
2 Hunger 1969–1970: 15; Hunger 1978: ii.324–5.
3 Mango 1975: 16–18.
4 Kazhdan 1995: 1–12; Cutler 1995: 203–14; For literary innovations in the eleventh and twelfth centuries see: Kazhdan and Epstein 1985: 83–6, 133–41.
5 See Reinsch 2010b: 56–61, for an overview of tradition and innovation in Byzantine literature. See Scott 1981: 61–74; Runciman 1995: 59–66; Spanos 2014: 44–5 for innovation in historiography. For a particular tenth-century example see: Kaldellis 2013: 35–52, on the literary innovations found in the lost source which recounts the campaign of Tzimiskes in 971.
6 Cutler 1995: 208–9; Nilsson 2010: 195, 207–8. Some case studies include originality in the comments of Photios and the *TNO*, see: Croke 2006: 59–70; McGeer 1991: 129–38. For other examples see the works of Reinsch 2010a: 23–32; Miller 1976: 385–95.
7 Hinterberger 2010: 187–203.
8 Nilsson 2010: 203–5.
9 This was acknowledged in passing by Dain 1937: 44, but was never studied in detail.
10 The way in which authors of Byzantine military manuals updated and adapted Onasander is studied in detail by Chatzelis, 'The ideal general and the impact of Onasander and rhetoric on Byzantine military manuals' in a volume edited by Dragana Dimitrijević under the tentative title *Reshaping the Classical Tradition in Byzantium: Texts and Contexts,* to appear in 2019.
11 Onasander, 37.3: 'τὸ μὴ πιστὸν διὰ τὸ ἀπεχθές' (trans. Oldfather and Oldfather, p. 495).
12 'τὸ μὴ πιστὸν διὰ τὸ τοῦ ἤθους βαρβαρικόν' (trans. Chatzelis and Harris, p. 35).
13 Several works are attributed to him, including a homily to Romanos I for the peace treaty of 927, and the imperial correspondence in the reign of Romanos I. For Daphnopates see: TC, 470; Jenkins 1966: 301–2; Stavridou-Zafraka 1976: 351–5; Dujčev 1978: 250–1; Darrouzès and Westerink 1978: 1–26; Flusin 2001: 48–50; Kazhdan 2006: 152–7; Chernoglazov 2013: 623–31.
14 Daphnopates, 85: 'τελείας καὶ ὀρθοτόμου πίστεως ὧν ἐραστης, διὰ τί μὴ πρὸς εἰρηνικάς καὶ ἀδιαλύτους συμβιβάσεις συνέρχῃ'; 'μὴ χείρονες βαρβάρων τῇ διαθέσει φανῶμεν, οἵτινες κἂν ὁπωσοῦν μεθ' ἡμῶν εἰρηνεύοντες καὶ ἀλλαγήν αἰχμαλώτων ποιοῦντες, ὡς μὴ τελείας ὄντες πίστεως, οὐδὲ τελείαν εἰρηνην ἀσπάζονται' (my trans. from the Greek).

For the use of the word barbarian in the Byzantine context see: *ODB*: i.252–3; Lechner 1955: 74–124, 69–106, in connection with religion. For Arabs as barbarians see also: Christides 1969: 319–24.

15 Stern 1950: 245–6; Lev 1995: 191–2.

16 *ST*, 1.7: 'Σώφρονα δε, ἵνα μὴ ταῖς φυσικαῖς ἀνθελκόμενος ἡδοναῖς τὴν περὶ τῶν μεγίστων ἀπολείπῃ φροντίδα' (trans. Chatzelis and Harris, p. 22); Onasander 1.2: 'σώφρονα δε, ἵνα μὴ ταῖς φυσικαῖς ἀνθελκόμενος ἡδοναῖς ἀπολείπῃ τὴν ὑπὲρ τῶν μεγίστων φροντίδα'.

17 Basil I, *Hortatory Chapters*, 1.11.

18 See for instance, Dennis 1993: 113–17; Strässle 2004: 121–4; McGuckin 2011–2012: 29–44. For the aspects of God granting victory through iconography see: Nelson 2011–2012: 162–92.

19 (trans. Chatzelis and Harris, p. 22); Stobaeus, 3.17.17. See also: Kaldellis 2007: 13–188, for a discussion of Byzantium and the reception of its classical past.

20 *ST*, 1.4.

21 *ST*, 1.5 (trans. Chatzelis and Harris, p. 22).

22 Kolias 2007: 320–5; Haldon 1999: 84, 148; 2014: 130; Ahrweiler 1960: 37–45.

23 *ST*, 2, 36; *LT*, 4.1, 11.9, 20.209. For such passages, see Kolias 2007: 323–5; Dagron and Mihăescu 1986: 260–74.

24 The protagonist is changed: *ST*, 80.5 with *Hypothesis*, 9.3 and Polyaenus, 5.33.3; *ST*, 83.1 with *Hypothesis*, 7.3 and Polyaenus, 5.33.6; *ST*, 85.2 with *Hypothesis*, 44.1 and Polyaenus, 3.9.57; *ST*, 86.5 with *Hypothesis*, 14.22 and Polyaenus 7.35. The stratagem is changed: *ST*, 79.2 with *Hypothesis*, 3.5 and Polyaenus, 4.3.32. For more examples and commentary see Chatzelis and Harris 2017: 140–50; Krentz and Wheeler 1994: xxii–xxiii; Dain 1931: 323–33.

25 (trans. Chatzelis and Harris, p. 94).

26 *MS*, 8.2.99; *LT*, 17.54, 18.129; Kolias 1989: 475–6; Haldon 2014: 323. While the use of poison is very limited in *LT* and absent in the *PM*, it does appear in the *De Obsidione*, 82–3 in a way reminiscent of *MS*.

27 *ST*, 65; *Apparatus Bellicus*, 36.

28 *ST*, 99.5; Krentz and Wheeler 1994: xxiii; Rance 2007a: 729–33.

29 See Chapter 6, pp. 88–92.

30 See for example, *ST*, 77.1; 76.11, 13; 97.2; 64; 77.1–2; 95.5; 102.3; Krentz and Wheeler 1994: xxii–xxiii; Meulder 2003: 445–66; Chatzelis and Harris 2017: 140–50.

31 Berger 2013: 247–58.

32 *ST*, 86.5; *Hypothesis*, 14.22; Polyaenus, 7.35.1.

33 For more information on promotional literature see Chapter 6, pp. 97–9. For the latest study on the relationship of the capital with the provincial aristocracy see Andriollo 2017.

34 On heroic ideals see: Holmes 2005: 240–98; Kazhdan 2006: 273–94; Stouraitis 2009: 114–19; Neville 2012: 2–27, 89–103, 121–38; Sinclair 2012: 319–79; Andriollo 2014: 126–38. Kaldellis 2007: 24. For the early Byzantine empire, see Steward 2016. For single combat see Kyriakidis 2016a: 114–36.

35 Skylitzes, 337 (trans. Wortley, p. 319); Holmes 2005: 269–72.

36 *LT*, 14.19; 20.190 (trans. Dennis, pp. 301, 605).

37 *ST*, 1.11 (trans. Chatzelis and Harris, pp. 22–3); Isocrates, *To Demonicus*, 43.

38 Kekaumenos 2 [trans. Roueché (online publication)].

39 Bryennios, 4.5; Neville 2012: 100–3, 126–8.

40 Skylitzes, 319–26; Sinclair 2012: 320–7; Holmes 2005: 289–98; Stouraitis 2009: 114–19.

41 TC (b), 3.22; Skylitzes, 67.

42 Arrian, 3.10 (trans. Brunt, p. i. 253).

43 *ST*, 48.7: Εἰδέναι δὲ χρὴ πρὸ παντὸς ὡς ἐν ἀδυναμίᾳ καὶ ὀλιγότητι στατοῦ αἱ νυχτομαχίαι ἐπενοήθησαν· ἀξιομάχου γὰρ ὄντος στατεύματος καὶ τὸ νικᾶν ἐντεῦθεν ἐφύβριστον καὶ οὐδενὸς ἄξιον (trans. Chatzelis and Harris, p. 81).

44 See Chapter 6, pp. 116–8.
45 Dagron and Mihăescu 1986: 139–44; McGeer 1995a: 191–5; Strässle 2006: 51–7.
46 See Chapters 1 and 3.
47 Haldon 1999: 149–65, 176–81; Dagron and Mihăescu 1986: 215–57, 275–87.
48 *MS*, 9.3.87–91, 12.20B.12–52; *LT* 9.37, 58.
49 *ST*, 52, 22.11; 40.7.
50 Haldon 1999: 217–20; McGeer 1988: 135–45.
51 See Chapter 3, pp. 43–4.
52 See for instance, Sekunda 2008: 339–43.
53 For the use of antiquarian terms to justify innovation see: Chapter 6, pp. 88–92.
54 *ST*, 38.6–7.
55 *DV*, 3.3, 20.73; *DRM*, 1.10–20, 35, 5.12–18, 6.1–5,58–60; *PM*, 1.51–62, 82–7, 95–7;
    Dagron and Mihăescu 1986: 192, n.42; McGeer 1995a: 208–9.
56 *ST*, 47.16; McGeer 1988: 135–45; 1995a: 209–11; Haldon 1999: 218.
57 *LT* 6.27, 9.71, 11.22, 19.14–16, 69.
58 Haldon 1975: 33; 2014: 202. For the debate about the *menavlion* see: McGeer 1986:
    53–8; Kolias 1988: 194–5; Anastasiadis 1994: 1–10; Dawson 2007: 7–8; Grotowski
    2010: 320–3.
59 For more details, see Chapter 7, p. 158.
60 Scholarship on the Roman heavy cavalry is vast. See, among others, Maspero 1912:
    105f; Eadie 1967: 161–73; Gamber 1968: 7–44; Diethart and Dintsis 1984: 67–84;
    Speidel 1984: 151–6; Dixon and Southern 1992; Mielczarek 1993; Negin 1998: 67–75.
61 The term *kataphraktos* appears in the *PS*, and *LT*, but only to denote the more heavily
    armed, see: *PS*, 35.19–20; *LT* 6.25–27, 15.9, 18.143,147, 19.14,73; *ODB*: iii.1114.
62 Haldon 1999: 219–20; Rance 2005: 552; 2008: 355–7; Trombley 2007: 347.
63 Procopius, 1.1.12–14, 5.27.4–5, 4.10.8–11, 6.1.29–34; Agathias, 1.2.5–8. 2.14.3; The-
    ophylact Simocatta, 6.2.3, 8.2.10; Corippus, 6.638.
64 *PS*, 17; *MS*, 1.2.35–9, 1.2.6–10; Jones 1964: 665–7; Haldon 1984: 102; 2002: 68; Bivar
    1972: 277–8.
65 See for instance: Ahrweiler 1960: 16; Dagron and Mihăescu 1986: 245; Haldon 1999:
    199–220; McGeer 1995a: 184–8; 273–4; 303–10; Trombley 2007: 347.
66 Kolias 1993a: 21–36; Wojnowski 2005: 7–20; 2012: 195–220.
67 The theory of Wojnoski cited earlier has some weakness. Wojnoski argued that the terms
    cataphracti, clibanari and *kataphraktoi* were solely used to describe armoured riders
    and horses of elite status. Since he accepts that the archer-lancer cavalry was the norm
    in the sixth to ninth century, he suggests that the term *kataphraktoi* was not used to
    describe them. This is not always the case though. As is highlighted by Jones 1964: 626,
    653, 661–3; Fear 2008: 450–1, some of the *cataphracti* of the late antiquity were of
    limitanei status and therefore did not belong to the elite of the army. Additionally, Kaegi
    1964a: 98–9; Haldon 1999: 195, 216; Rance 2008: 356–8 have expressed strong doubts
    as to whether the archer-lancer cavalry should be taken for granted that it was part of
    the normal fighting force. Last but not least, most of the time Wojnoski assumes that the
    appearance of the word *kataphraktos* is directly connected with a specific type of Byz-
    antine heavy cavalry. He therefore overlooks the fact that the word *kataphraktos* was
    also used to describe a piece of armour or heavy armed troops in general, both infantry
    and cavalry, and so its presence is not always proof of continuation in equipment or
    tactics.
68 Some examples include: *PS*, 35.20; *LT*. 6.163–172, 15.52–54, 18.775, 816, 19.86, 404;
    Psellos, *Chronographia*, 6.87; Anna Komnene, 5.6.4.4, 15.6.4.14, 15.6.17.3; Attaleiates,
    2.62–63.24.
69 See Chapter 3, pp. 44–6.
70 *ST*, 46.6–8; McGeer 1995a: 286–9; Haldon 1999: 220.
71 *LT*, 18.26; *ST*, 46.26; Haldon 2014: 337.

72 *ST*, 45.10.
73 *ST*, 47.1 (trans. Chatzelis and Harris, p. 75); McGeer 1995a: 262.
74 *MS*, 7B.11.42–52; *LT*, 14.20; McGeer 1995a: 259–60; Haldon 1999: 219.
75 *ST*, 47; McGeer 1988: 135–45; 1995a: 257–80; Haldon 1999: 217–20.
76 *LT* 18.113–14 (trans. Dennis p. 479).
77 Haldon 2014: 360.
78 See Chapter 7, pp. 158–60.
79 *ST*, 45.
80 *LT*, 4.64–76.
81 See Chapter 6, p. 133.
82 *LT*, 12.20–5; *ST*, 46.
83 *ST*, 46.17, 19; Dagron and Mihăescu 1986: 59, n.4; McGeer 1995a: 283–4; Sullivan 2010a: 155.
84 See Chapter 7, p. 156.
85 For an overview see Haldon 2006: 290–7 and Sullivan 2000: 231.
86 Heron of Byzantium, *Parangelmata Poliorketika*, 49.20–25; *De Obsidione*, 64.113. There is a surviving representation of this device in the Vaticanus gr. 1605, folio 36, see: Sullivan 2000: fig.22.
87 *ST*, 53.8 (trans. Chatzelis and Harris, p. 88).
88 *LT*, 19.64.
89 *LT*, 5.3; 6.21; *ST*, 38.1–5; Kolias 1989: 103–9; Dawson 2007: 2–5; Sullivan 2010a: 155. For tenth-century Byzantine equipment see also: Dawson 2002: 81–90; 1999: 38–50; 2001–2002: 89–95; Grotowski 2010; Haldon 1975: 11–46; 2002: 65–79; 2014; Tsurtsumia 2011b: 65–99.
90 *LT*, 6.22; *ST*, 38.8–11.
91 *LT* 6.1–7, 6.23, 7.41; *ST*, 39.1–6; *PM* 3.26–46; Grotowski 2010: 367.

6 The *Sylloge Tacticorum* and its practical use

The *ST* belongs to the literary genre which is usually described as military manuals or treatises. As with almost any other treatise of that kind, the *ST* contains a great deal of information which can be traced to older sources, some of which date back to antiquity. Therefore, with such a degree of antiquarianism in its contents, it is only natural to ask what the purpose of the *ST* was. It could have been for practical use on the battlefield or just a good book to read by the fire, a literary work which mostly aimed at preserving the older tradition.[1] Before this issue can be resolved in the case of the *ST*, one needs to consider the bigger picture and examine the purpose of military manuals in general. Were military manuals antiquarian literary projects of more interest to scholars than generals? While the individual treatises within the genre differ widely in some respects, the broader context will provide an important entry to an assessment of the purpose of the *ST*.[2]

Military manuals as antiquarian literary projects

The presence of antiquarian material appears awkward to our modern perception of what a practical manual or handbook should look like. The fact that a great deal of the material of Byzantine military manuals derives from classical and/or older Byzantine works is the main reason why these treatises have been labelled 'works whose primary aim was to preserve the ancient knowledge', 'more literary than technical', 'theoretical rather than practical' and 'traditional rather than innovative'.[3] Although these comments are usually moderated when studying specific military treatises, that is not always the case. For instance, we have seen that the *ST* is sometimes regarded as a purely literary manual, copied with very little adaptation from its sources, and as a source which has little relevance to tenth-century Byzantine warfare.[4] Thus, military manuals are envisaged mainly as literary projects which mostly interested scholars and politicians. From this point of view, the purpose behind the production of these works seems to vary. Their authors may have aimed to preserve the classical tradition, to present themselves as its protectors and continuators, to seek advancement or legitimacy or even to experiment linguistically.

Nevertheless, when we argue that the presence of antiquarian material is enough to qualify a military manual as non-practical, we tend to judge Byzantium through a contemporary way of thinking. This approach overlooks the important role that the past played in Byzantium. We have already seen that the presence of older works in military treatises can be adequately explained in the context of high Byzantine literature and education, as well as the relationship between patron and author.[5] Before we put a certain label to a work, it is necessary to take into account how the Byzantines used the past to educate themselves and to wonder whether the past enjoyed more credibility than the present.

Byzantine society was averse to radical or sudden change; pure originality was seen as something negative, and the word innovation, *kainotomia*, was used to negatively describe a breach with the tradition. But as is the case with most societies, Byzantium did not remain stagnant, for it was quite receptive of gradual innovation.[6] In this light, certain ancient or older practices became authoritative, and they were justified by the mere fact that they belonged to the past.[7] In some genres of Byzantine literature, older established insight seemed to be more credible than contemporary observation, and there were cases in which the authority of the past was so strong that a tradition had to be fabricated in order for one to advance his case.[8] Thus, anyone who created something new had to justify their innovation by putting it in a context of older material, which would highlight a gradual rather than a radical change and which would connect the innovation with the authority of the tradition.

Influenced as they were by this approach, the Byzantines saw both ancient military manuals and history as a credible source of educating themselves about contemporary matters.[9] For example, the author of the *ST* twice states in his manual that 'the sufferings of the ancients should be a lesson to the contemporaries'.[10] John Doukas and Alexios I are recorded as having read the works of Aelian and Apollodorus.[11] Constantine VII especially stresses that the stratagems of Polyaenus and other historical works must be part of the baggage train during a campaign, while Theophylact Simocatta states that Philippicus, 'who was very fond of learning, drew his military knowledge from the experts of the past', and thus imitated Scipio.[12] Furthermore, Julian (361–363) reports that he educated himself and his officers with older writings so as to use them as exemplars and to avoid repeating the same mistakes.[13] Ammianus Marcellinus seems to confirm this mentality, as he states that Julian, being acquainted with historical deeds, was inspired to attack Pirisobora, much like Scipio did in the case of Carthage.[14] Leo VI encourages his readers to look for information and ideas regarding siege matters in historical works.[15] The use of ancient exemplars and the need for the Byzantines to emulate ancient prominent figures seem to be well underlined by the author of the *ST*. More specifically he states that a general will never achieve something noteworthy 'if he does not boast the wisdom and judgment of Solomon, the strength of Samson and Hercules, the tactical experience and ability of Cyrus and Alexander and the fortune of Caesar'.[16] The case is no different in Kekaumenos, who underlines his advice with figures of the past such as Pyrrhus, Scipio, Hannibal or Belisarius.[17]

By taking these factors into account, we can further understand and explain the purpose that antiquarian material played in military treatises: it was considered to be an exemplar more credible than contemporary observation. This explains why authors such as Maurice, Nikephoros II and Nikephoros Ouranos, who had first-hand military experience and could have highlighted their personal deeds, chose not to do it, but preferred the credibility of sticking to their older sources, copying or paraphrasing them. Therefore, it seems that whenever an older author had already referred to something that a contemporary military writer wanted to discuss, the latter preferred the credibility of the older model with minor changes or additions. In this way the author not only showed that he was familiar with the older works but also gained credibility. He justified his innovations by presenting them as a slight addition to established practices or explicitly referred to the past to explain why certain practices were no longer used in his time.

Of course, every author and work were different. Some manuals seem more practical than others, and some authors appear to be less or more conservative. Be that as it may, there are numerous authors of military manuals who seem to support this approach, since their personal comments seem to imply an 'undeclared war' between the credibility of the past and the need to include innovative material. For instance, Urbicius reports in his *Epitideuma* that he was asked to present a formation that will ensure the safety of the army but hopes that he 'will be excused' since his battle array is his own invention and not connected to the ancient tradition. Towards the end of his work he makes reference to the same issue again, hoping that his formation will not 'be despised just because it has been invented for the first time and has not yet demonstrated its usefulness in practice'.[18]

Maurice reports in his *Strategikon* that he will draw on both ancient authorities and his 'limited experience'. It could be argued that Maurice employed a *topos* here: he deliberately tried to humble the value of contemporary experience as opposed to the well-established ancient authorities. This seems to be further supported by the fact that he continues to state quite apologetically that he 'has no pretence of breaking new ground or of trying to improve upon the ancients', for he recognizes them as necessary.[19] In addition, Leo VI felt the need to explain that first he tried to find information for naval warfare in the older tacticians and when he failed to do so, he consulted contemporary commanders.[20] Similarly, Constantine VII explains that he did a lot of research to find sources which deal with the matters of an imperial expedition. But he was only able to find the work of Katakylas, which was not to his liking; therefore, he altered it, probably with the help of individuals who had some personal experience.[21] Constantine VII fabricated authority for his material, stating that ninth-century practices were undertaken by Constantine the Great (324–337) and Julius Caesar (d. 44 BC).[22] The author of the *Geoponika* states that he does not regard all the evidence from the authors of antiquity he has cited earlier as credible and true, but he explains that he added their views so that he may not be regarded as having omitted the words of the ancients.[23]

Moreover, the author of the *ST* reports that it is necessary for those who write about tactics to cover both ancient and contemporary practices, so as for his reader to become well versed in the deeds of both ancient and contemporary generals because that is how a military manual becomes effective.[24] It seems that the author of the *ST* placed references to antiquarian material carefully in order to justify some of his major innovations. For example, before he deals with the new hollow-square formation, he explicitly refers to its originality. Right after the discussion of the square formation, however, it seems as if the author felt the need to link the new with the old, and so he explains that Aelian and Polybius had proposed to increase or decrease the number of intervals according to the situation.[25]

Likewise, perhaps to justify the introduction of new troops such as the *peltastai* and the *kataphraktoi*, the author of the *ST* first decided to present the armament of their Hellenistic counterparts, which is given in Byzantine terminology.[26] Then the contemporary situation is presented, and it comes as no surprise that Byzantine troops are also referred as *peltastai* and *kataphraktoi*, perhaps to further imply that there was no rigid break with the tradition.[27] In addition, even Nikephoros II Phokas, an author who was quite original and treated his sources more freely, felt the need to explain why he differentiated himself from the Hellenistic phalanx. More specifically, Nikephoros II explains that so deep a formation was out of use in his day, since in contrast to ancient practices, the Arabs have reduced their depth, thus rendering the Hellenistic formation obsolete.[28]

However, other authors were not so conservative; instead, they emphasized that personal experience and creativity should also play a part in warfare and strategy. For instance, Nikephoros Ouranos used numerous ancient works, and the only surviving title of his treatise in the *Constantinopolitanus Gr.* 36 cites no fewer than eighteen ancient authorities. Nevertheless, he differentiates himself from ancient siege practices, explaining that in contrast to classical authors who preferred siege engines, his experience has shown that undermining the foundations is the most effective way to capture a fortification, and he continues to note that he will only refer to methods which are currently in use.[29] Additionally, while the author of the *DV* cites older military manuals, he reports that he will also write according to his own experience.[30] To further underline his approach, he clearly states that

> All this we are setting forth as experience teaches. It is up to you to apply it to circumstances at the urgent needs of the time. For tradition alone does not do it, but it must be reinforced by the assistance of God, and only then the outcome of the battle shall be assured.[31]

Katakalon Kekaumenos seems to support this approach; on the one hand he advised his readers to read ancient manuals, but on the other he went further to report that:

> Devise what you need yourself as well, not only what you have learned and heard of from the ancients, but think up other new things, which are within the capacity of human nature to invent. Don't say; 'It wasn't handed down to us by the ancients'. I tell you that human nature possesses innate cunning and wisdom; just as those very ancients invented their devices, you make your own discovery yourself and win a victory. For certainly they were men just like you.[32]

Last but not least, we should not be too keen to associate antiquarian material directly with outdated practices. We ought to take into account that the basic practises of warfare remained largely unchanged before the invention of gunpowder. This is not to say that nothing has changed, but the very bases of leadership, siege warfare and battle formations were not altered. Be that as it may, Onasander's work was still very relevant to any tenth-century Byzantine general.[33] Moreover, the Byzantines could find a great deal of useful information from older tacticians or historians, and it is no surprise that tenth-century siege manuals specifically cite and use them as their sources, sometimes revising and updating them.[34] Finally, while armament, formations and the roles of infantry and cavalry had changed, the commander could still look for ideas in the ancient formations of Aelian. It is no coincidence that Katakalon Kekaumenos, who was a veteran, urged his readers to do so, and after all, Aelian also appears in a number of Byzantine manuals such as the *PS*, *LT*, the *ST* and the *TNO*, which draw selectively on him, either directly or indirectly.[35]

Apart from antiquarianism, political promotion and propaganda are other factors found in military manuals which could qualify them mostly as literary projects. As is the case with tenth- and eleventh-century compilation literature in general, one may perceive the purpose of military manuals as mostly aiming to highlight legitimacy or to persuade readers that their patron or author was the most suitable person to gain authority or offices.[36]

Such markers of legitimacy and political propaganda are indeed evident in military manuals, a good example being *LT*. Leo VI refers there to his father and predecessor, Basil I, as a prudent and effective general who successfully and safely crossed a river with his army by implementing the instructions that Leo VI gives in his manual. Basil I is also described as a benefactor of the Bulgarians who were 'crude' and 'barbarous' before they were baptized by him.[37] Furthermore, it is possible that Constantine VII might have shown such an interest in writing or collecting military manuals, so as to compensate for the fact that he had no military experience, thus trying to present himself as somebody who had active authority over the army. This argument could also be applied to Leo VI, who, unlike his father, was also an 'armchair' general.[38] In addition, other military manuals make an explicit effort to stress the deeds of specific individuals. Such propaganda is visible in the *DV*. This treatise seems to have been compiled under the orders of Nikephoros II Phokas by one of his officers who was closely associated with him.[39] In this manual we find direct or indirect references to successful operations of generals who belonged to the Phokas family, such as Leo, Bardas and Nikephoros Phokas, brother, father and grandfather of Nikephoros II, respectively, while the deeds of other generals are passed over in silence.[40] This has led certain scholars to argue that the primary purpose of the *DV* was political propaganda or even political criticism.[41] Finally, a dedication poem placed before the preface of Basil Lekapenos' *Naumachica* has also been interpreted as a case of political progression. The preface records the victories of Basil against the Arabs, stating that he will also enjoy similar victories in the sea, which can be interpreted as aiming to persuade Romanos II that Basil Lekapenos was a more suitable candidate than Joseph Bringas for political promotion before the Cretan expedition.[42]

While this 'legitimacy and political progression' approach fits well into the tenth-century context, and there could be various motives behind a work, the evidence found in the military manuals themselves does not always clearly show that progression was the main factor behind their compilation. In addition, even if the author's ultimate motive was merely to promote himself, he could have done that, and perhaps better, by producing a practical manual. To begin with, the dedicatory poem in Basil's *Naumachica* could have merely been an attempt on the part of the author, who has been identified as a young and minor figure in the circle of Basil Lekapenos, to express his goodwill and praise to his patron.[43] This becomes more probable if we accept the latest view which suggests that the manual had little or no practical use, interpreting it more as a linguistic exercise or a scholarly game of antiquarianism.[44] Therefore, one might wonder why Basil Lekapenos chose to promote himself against his political opponent by being the patron of a manual which had no or little practical use.

With regard to the *DV* and *LT*, where reference to a certain family member or to a predecessor is found, these do not seem to be central or extensive. Basil I is mentioned only twice in *LT*, while the general Nikephoros Phokas, grandfather of the future Nikephoros II, is mentioned three times.[45] The fact that Nikephoros, who had no dynastic connection with Leo VI, is mentioned more frequently than Basil I seems to imply that the main focus of Leo VI was practical, namely to provide his reader with recent examples of loyal generals who acted according to the advice of *LT*. While the reference to Basil I also served to underline the legitimacy of Leo VI, this does not seem to be among the dominant reasons for the creation of *LT*. As far as the *DV* is concerned, political propaganda can be mostly seen in the preface, rather than in the main body of the work. Whether it also expressed political criticism is something whose relevance depends on its dating.[46]

Consequently, while legitimacy and praise of a certain family are present in manuals, they do not appear to have been the major factors behind their compilation. Legitimacy and praise seem to be present in the customary context in which the author would praise his patron, imply his legitimacy or make specific reference to loyal family members. By the same standards we cannot explicitly argue that the *De Thematibus*, which refers kindly to Romanos I, was written with a main purpose of praising Romanos I, since we know that Constantine VII was deeply hostile towards him. Similarly, the references to Basil I and Leo VI and the negative ones to Romanos I cannot explicitly mean that the main purpose of the *DAI* and the *DC* was to praise or to condemn these figures. It cannot be overlooked, of course, that propaganda was important and practised through literary works when the opportunity presented itself.[47] Textual evidence, however, seems to suggest that this propaganda was sort of a 'happy accident' which mostly highlighted a situation that already existed, rather than aiming to create a new equilibrium or being a major factor behind the drafting of manuals.

This line of reasoning is also supported by the fact that some manuals contain no references to previous emperors or families. For instance, the *De Obsidione*

Toleranda, the *Syntaxis Armatorum Quadrata*, the *DRM* or the *Memorandum* not only lack references to specific contemporary individuals, but they seem to have been written by ghost authors; a fact which seems to remove political progression, influence or propaganda as a major factor in their composition. The *ST* itself fits well into that pattern. In addition, even though a number of military treatises are not anonymous, political propaganda or highlighting personal deeds to achieve political progression is usually not obvious. For example, Maurice, the author of the *Strategikon*, does not underline any of his achievements as a general, neither does he praise or criticize contemporary individuals in his manual.[48] Similarly, in the *PM*, or at least in the version that is extant today,[49] Nikephoros Phokas, an experienced and successful general himself, did not feel it necessary to include any of his family or personal deeds, even though some of the latter, especially the battle outside of Tarsus according to the account of Leo the Deacon, seem highly relevant.[50] What is more, any other form of political propaganda against or in favour of other individuals is equally absent. An exception might be Syrianos Magister, the author of the *PS*, who has been identified by some as a person who might have had first-hand military engineer experience and who clearly states that he can improve Apollodorus and the ancient siege treatises.[51] His reference to Belisarius, however, does not seem to serve any political end, if we accept the latest studies that the treatise was written in the ninth century.[52]

Going into the eleventh century the picture does not seem to alter. One could suggest that one of the primary reasons for the compilation of the *TNO* was to advance the author politically in the context of peer rivalry between Byzantine generals under Basil II.[53] This may well be true, but it does not come across from the text of the *TNO* itself. The part of the treatise that has already been edited at least does not include any hints of political support towards any particular figure, and there is also a lack of reference concerning the deeds of Nikephoros Ouranos himself, even though it is likely that the work was written after Nikephoros Ouranos had achieved his greatest military deeds.[54] If progression was the main aim of Nikephoros Ouranos, he could have better achieved his goal through such references. For example, when he discusses the crossing of rivers in his manual, he could have mentioned his personal experience, when, in 997, he crossed Sperchios River and suddenly attacked the camp of Samuel.[55]

Military manuals as practical handbooks

Having discussed the factors which contradict our modern perception of what a practical manual should look like, it will be worthwhile to go deeper into the other side of the argument, because there is more direct evidence in favour of the practical use of at least certain manuals. This evidence can be grouped into four main categories, and examples from the *ST* can be seen in all four.

The first concerns how the authors of military manuals saw their works. Judging from their comments, they regarded their treatises as practical handbooks rather than purely literary works. Despite the fact that the author of the *ST* records that his book would be effective if it covered both ancient and contemporary practices, he reports that he will put emphasis on what is contemporary.[56] Maurice begins his work by stating that he will focus on practicality and that his book is 'an elementary handbook for generalship'.[57] The same message appears indirectly in *LT*. Leo VI had written his book in a form of imperial legislation, always addressing it to the generals, which seems to imply that he saw it as a way of legislating for warfare and therefore probably expected his advice to be followed as an order. What is more, in the epilogue of the work we read that the book was compiled in response to the Arab threat, but this might have been absent from the original version and could have only been added in the version known from *Laurentianus Plut. 55.4*.[58] The author of the *Poliorketika Parangelmata* states that he is writing to present the ancient siege engines, but also to update and make them relevant to his time, since according to him, his target audience is the military commanders, and it is they who will benefit from his work.[59] A hint towards practical use is also given in the *DV*. Although the author cited information from his personal experience and that of other contemporary commanders, he felt the need to apologize that his tactics 'might not find much application in the Eastern regions at the present time', due to the fact that the Byzantines were mainly on the offensive.[60]

Nikephoros Ouranos implies the practicality of his work in a different way. He states that 'it is inappropriate for some [devices] to be described to prevent their becoming known to the enemy'.[61] It seems, therefore, that Ouranos regarded Byzantine manuals as practical handbooks which were to be carried in the battlefield, and as a result implies that some of them were deliberately generic for fear of being stolen by the enemy. Ouranos' concern does not seem to be completely unfounded. Constantine VII reports that military manuals were intended to be part of the imperial baggage train and could therefore end up in the hands of the enemy if the latter was looted.[62]

We are aware of some cases in which the Arabs had at their disposal material deriving from treatises of antiquity or even from Byzantium. The oldest was compiled by al-Harthamī, written for the Arab caliph al-Ma'mūn (813–833), but unfortunately it is only known to us in an abridged form which is preserved in much later manuscripts. The original work appears to have been called 'Al-Ḥiyāl' (Stratagems) and seems to include passages which originally derived from Aelian, Arrian and Onasander. Some of its contents seems to have been very similar to those of Byzantine military manuals, covering aspects such as the qualities of a good general, espionage, ambushes and night attacks, formations, sieges, camps and marches, but it might be that its material rather reflected classical practices than contemporary Abbasid ones.[63] Another work, an anonymous treatise on military organization and stratagems, was written in the late ninth or early tenth century, the author of which claims to have been inspired by an ancient Greek passage he found himself, while another fourteenth-century Arab writer, al-Aqsarā'ī, drew heavily upon Aelian.[64] The final case concerns *LT* which was partly or wholly at the disposal of the fourteenth-century Arab war writer Ibn Mankali, who paraphrased and used parts of it in his own manual. It is unclear, however, whether he translated the work himself or used an existing translation.[65] Regardless of the process, in the eyes of Ibn Mankali, who originated from a military family and was also a high-ranking military officer himself, the treatise of Leo VI was a manual 'written for the use of the Byzantines in their warfare against Muslims' and 'contains great military benefits'.[66] He also comments that it is 'useful for anyone who is engaged in fighting the enemy on land or at sea', which seems to imply that he regarded it as a practical handbook.[67]

The second group of arguments in favour of the practical use of manuals is connected with the concept that these treatises were written in order to tackle new dangers and respond to current threats or needs.[68] This idea is dependent on original material that is found in certain manuals, which provides us with an updated version of tactics and/or equipment. This is especially true for the manuals of the tenth century, where a gradual evolution can be observed from *LT* to the *ST*, to the *PM* and the *DRM* and finally to the *TNO*.[69] From a slightly different perspective one may argue that the original material of military manuals touched upon both current practical and intellectual problems. This is especially evident in *LT* where the morale of the army plays a central role and where Leo VI attempts to revive it using Christian theology. Thus, similarly to the ninth-century intellectual context of anti-Muslim polemic, Leo VI underlined the superiority of Christianity and created a Byzantine philosophy of warfare.[70] These ideas do not only concern military manuals, they also extend to other treatises. For instance, the *DC* is no longer regarded a manual of outdated practices, but as playing a central role in the restoration or renovation of ceremonial practices in the time of the Macedonian dynasty.[71]

In the previous chapter, we discussed the military innovations that are found in the *ST* and how these fit the new military context of its time, as well as how its author seems to have been aware of the new challenges and roles the army had to fulfil.[72] Other examples include the original material of *MS* which seems to highlight the adaptation of the Byzantine army to fight against mobile nomads, Slavs and Persians.[73] Some scholars have gone further than that and by comparing the original material of *MS* to the historical narratives, they have concluded that certain generals, such as Herakleios, were familiar with its instructions and practically applied them.[74] Later Byzantine manuals are no different. *LT* contains original material which is devoted to Arab–Byzantine warfare, and apart from new tactics, it also records shifts in attitudes and perceptions of war.[75] The same applies to the *PM*, *DV* and *TNO*, which feature an evolution of Byzantine strategy, armament and tactics, as well as equipment, and contain original material which is in practical accordance with the situation on the Arab–Byzantine borders.[76] Last but not least, similar observations can be made for the manuals which focus on the sphere of siege warfare. For instance, the *Poliorketika Parangelmata* and the *De Obsidione Toleranda* present innovative material, parallel examples of which can be found in historical narratives.[77]

The third argument that wants manuals to have played a practical role has to do with how the Byzantine court functioned. In the Byzantine world there was no military academy as such. Therefore, military knowledge could only be acquired by limited and specific means. The most obvious way would be to spend some time in the army, to be promoted to an officer and acquire first-hand experience. Another possible way would have been for someone to be part of a military family and thus acquire experience either by being instructed by a senior officer or even campaigning with him to observe. According to *TC*, that is how Basil I chose to instruct Constantine, his intended heir, by bringing him along on expeditions, 'to (. . .) be his teacher in tactics'.[78] The third way, which was the quickest and least risky, was to learn about military matters by reading. The writings which somebody could have consulted may have well varied. We have already seen that some military commanders read ancient histories in order to educate themselves about war, but historical texts did not always provide detailed coverage on all aspects of warfare.

Another possible source of learning could have been to read 'promotional literature' of the military aristocracy, such as biographies of famous generals and families like the Kourkouai or Phokades. Although these sources are no longer extant, it can be speculated that they probably included a great deal of military information. However, this seems to have been a genre that did not start to be established until the middle of the tenth century, so it does not seem to have affected matters much by the time the *ST* was compiled.[79] An additional solution, although undoubtedly quite an exclusive one, would have been for somebody to consult official army reports or correspondence. These military bulletins were sent from the field to Constantinople to inform about the course of events and results of battles. Nevertheless, even if somebody was in the position to have access to this material, it seems that records were not systematically kept, and thus probably these reports played little or no educational role.[80]

Finally, one could consult military manuals, and there are a number of primary sources which record their use in war or for educational purposes.[81] We have already seen that Constantine VII instructed that military manuals should be part of the imperial baggage and that Caesar John Doukas and Alexios I were recorded as being familiar with the writings of Aelian. According to Psellos, Basil II acted much as Kekaumenos and the *DV* proposes, drawing up his army by 'having read of some [formations] in the manuals and others by devising himself during the operations of war, the result of his own intuition'. Likewise, Psellos reports that Romanos IV (1068–1071) was envious of him because he 'was thoroughly conversant with the science of military tactics, that had made a complete study of everything pertaining to military formations, the building of war-machines, the capture of cities, and all the other things that a general has to consider'.[82] The author of the *Vita Basilii* seems to further imply the dominant role of military manuals in educating potential commanders. He states that 'were it possible for everyone to learn military science or art without study and considerable practice, authors of works on tactics who devote so much labour to this topic would be merely ranting senselessly'.[83] In addition, Nikephoros Bryennios records that Basil II, who was responsible for the education of Isaac and John Komnenos, appointed tutors so that among others the boys will 'study military treatises, so as to learn how to draw up a formation, array the files, how to pitch camp correctly and set up a palisaded encampment, and all the other things that the tactical manuals teach'.[84]

One may argue that such comments were promotional, aiming to present figures in an ideal light, but the mere fact that authors chose to promote figures in such a way implies that this was the way things were supposed to be done.[85] Nevertheless, the practical need for military manuals in Byzantine society is better highlighted by the fact that the main qualification to become a *strategos* or a military commander was to possess the full trust of the emperor. In the tenth-century context of powerful military families, the emperor wanted to make sure that no *thema* would revolt against him and that the person in charge of an expeditionary force would not attack Constantinople. Therefore, military experience and knowledge seemed to have been of secondary importance compared to loyalty. Under those circumstances the role of such manuals could prove invaluable. We possess a number of examples of how military operations resulted in disaster due to military blunders, which almost all military manuals warn against.

A popular example is that of Constantine Gongylios, a eunuch who enjoyed the trust of Constantine VII, holding the title of *patrikios*. Put in charge of the campaign to re-conquer Crete in 949, Gongylios failed to post sentries at night and secure his camp, with the result that the Byzantine force was slaughtered by an Arab night attack. Both John Skylitzes and Leo the Deacon describe him as somebody who had no experience with warfare.[86] Even if we argue that it was not entirely the fault of Gongylios, or that Leo the Deacon was biased against eunuchs,[87] the account of John Skylitzes seems to verify that some commanders lacked military experience.[88] For instance, Krateros, *strategos* of the theme of Kibyrrhaiote, committed a similar blunder in Crete during the reign of Michael II (820–829), and the case was no different in the reign of Basil I.[89] Basil was convinced to replace the more experienced commander in the east with Stypeiotes. Stypeiotes once again failed to post sentries and secure the Byzantine camp; thus, the Byzantine army was annihilated by a night attack, something which is also corroborated by Arab sources.[90] Stypeiotes is described as somebody who had no clear plan in mind and who lacked prudency and experience.[91]

Historical narratives and the practicality of military manuals

Except for educational practices there is also a final argument in favour of the practical use of military manuals. A number of historical narratives seem to provide similar information to the military treatises or describe Byzantine commanders acting in similar ways. This evidence has led a number of scholars to conclude that indeed certain generals were aware of them and applied their instructions on the battlefield. Such examples include Emperor Herakleios and *MS*; Leo the Deacon and the *PM*; and siege descriptions from historical narratives up to the *Alexiad* and the manuals of *ST*, *Parangelmata Poliorketika* and the *De Obsidione*.[92] However, the parallel information between the historical narratives and the military manuals is, of course, something a historian cannot readily accept as proof that the advice of the manuals was applied on the battlefield. This is especially true when somebody attempts to identify the practical use of the *ST* due to the fact that the major Byzantine sources for the tenth century were either written in the second half of the century or in the eleventh or even in the twelfth century. This creates a difficulty that not only concerns the way in which Byzantine historical narratives chose to describe warfare but also the fact that many of our historical narratives seem to have drawn on non-extant sources.[93]

Some of these lost sources were biographies or encomia of successful generals or battle reports written in the context of peer rivalry between powerful families.[94] For instance, we have evidence for the existence of a book that covered the military deeds of John Kourkouas and for others which were dedicated to Nikephoros II.[95] This military promotional literature can be problematic for two main reasons. First, it seems to have included praise for the figures it was dedicated to, and second its target group seems to have been the military aristocracy, namely men with military interests and experience who could have enjoyed reading about idealized generals who acted similarly to the treatises.[96] With regard to battle reports, although they could have been a more credible source, there is one point of concern, namely, that one might choose to distort actual events to his advantage, especially when we know that these reports were sometimes read publicly or even sent to foreign rulers.[97]

When we take into consideration these factors together with the militaristic context of elite Byzantine society from the tenth century onwards, we are faced with one major concern: Are the parallel events between the manuals and the historical narratives proof of the practicality of the military treatises, or have they been included in the histories as part of lost sources whose purpose was to present figures as ideal generals who played 'by the book'?[98] For example, Bardas Skleros is described on the battlefield by John Skylitzes and Michael Psellos as acting in accordance with the advice of the *PM* and *TNO*. Should we take this information at face value or perceive it as deriving from a pro-Skleros source which aimed at presenting him as an ideal general?[99] Similarly, it is rational to wonder whether the same issues apply to Leo the Deacon's narrative of Nikephoros II Phokas, Leo Phokas and John Tzimiskes, where fortified camps, night attacks and frontier warfare all reflect the advice of the *PM* and the *DV*.[100]

It would be better neither to reject nor to accept the practicality of the manuals on the basis of the parallel evidence between historical narratives and the treatises. It would be too easy for the practicality of the manuals to be either overestimated or underestimated. After all, these sources are not extant, and their exact contents or distortions can only be guessed at. The mere fact, however, that these lost sources seem to have chosen to portray the generals as playing it by the book, even if their evidence is entirely false, implies that the instructions of the manuals were not something marginal and entirely theoretical, but rather how the general was expected to act, at least ideally. This shows that they had been an integral part of the way military aristocracy thought and understood warfare.

Therefore, the degree of practicality of the *ST* can be estimated by comparing its instructions with the historical narratives of this era, while also considering the effect of these lost promotional sources. This can be achieved by approaching our sources more critically and comparing them not only with other types of Byzantine sources but also with Arab, Western and Armenian material. We will study the degree of practicality of the *ST* focusing on aspects and traits of generalship, rewarding and punishing soldiers, gathering intelligence, camps and night attacks, siege warfare, numbers units and formations, as well as guerrilla warfare.

Practical information 1: aspects and traits of generalship

It is no surprise that the *ST* is addressed directly to the general and that he is, to a great extent, the centre of its focus. The manual does not only advise the general on tactics and formations but also guides him on how to cultivate his character and stance on warfare. All these shape the image of the ideal general, and the vast majority of these passages derive from Onasander.[101] Although the work of Onasander was very old, it was still quite relevant, and with minor additions and adaptations became ever more so. Onasander's influence becomes more evident when we take into account that apart from the *ST*, extensive passages of his work can be found in other military manuals as well, such as *LT* and the *TNO*.[102] What makes the practicality of these passages difficult to estimate is that most of them are connected with moral characteristics which Byzantine historians would manipulate according to the situation to make a point to their readers.[103] These moral qualities occur in Byzantine histories and are highlighted in such a way so as to appear as a major factor in the outcome of a battle. Thus, historical narratives which draw on promotional literature seem to deliberately associate successful commanders with these moral characteristics, whereas disaster is blamed on the lack of such qualities.

A good example of adverse traits being employed by Byzantine historians to cast their subjects in a negative light and to create a moral cloak for defeats can be found in the narratives of Leo the Deacon and John Skylitzes. The *ST* describes the ideal general as 'prudent, so that he might not abandon his devotion to the most important [things], by being drawn to physical pleasures', because 'extravagant indulgences (. . .) do not allow for keeping vigilant in the most important matters'.[104] In this light, Leo the Deacon and Michael Psellos idealized Nikephoros II Phokas and Basil II, respectively, presenting them as men with 'a temperate disposition and not tempted by pleasures', who endured 'the rigours of winter and the heat of summer', as well as 'kept natural desires under stern control'.[105] In juxtaposition, a Byzantine defeat in Crete is blamed on Nikephoros Pastilas, and the fact that he was ambushed by the Arabs is explained by his depravity since he and his troops were absorbed by the riches of the countryside and 'indulged in indolence and luxury'. Leo the Deacon continues the story underlying his moral message with the prudent Nikephoros II, giving a fictional rhetorical speech on how victory will be assured if everybody refrains from physical pleasures.[106]

Similarly, contrary to the ideal general who ought to be 'firm so as not to be puffed up in times of victory', 'open to counsel' and 'not so stubborn, so that it is impossible for something better to be thought, considered and said by somebody else',[107] Manuel is described as 'hot-headed and self-willed' or as a man who ignored good advice.[108] Thus, Leo the Deacon and John Skylitzes justified the Byzantine defeat in Sicily (964–965) on the fact that Manuel was influenced by his early victories and insisted on pursuing the retreating enemy in difficult terrain. It is interesting that this imagery is not limited to the Byzantines. Miskawaihi and Ibn al-Athīr attribute the defeat of Sayf al-Dawla in 950 to his stubbornness and his failure to accept that the Tarsians possessed a better espionage system than him. Sayf ignored the intelligence of the Tarsians, who reported to him that the passage he intended to retreat through was already occupied. This accusation might not be entirely fair though, as there was distrust between Sayf and the Tarsians at that time.[109]

While in Byzantine historical narratives a stubborn and immoral general equalled a curse, a pious, prudent and modest one equalled a blessing.[110] Following advice originally found in Onasander, the author of the *ST* reports that 'when an essential piece of state business presses, in which the army has to serve the general with manual labour, the general himself should be the first to begin the work'.[111] In accordance with this advice, the history of Leo the Deacon and the *Vita Basilii* glorified Nikephoros II and Basil I by reporting that they eagerly loaded their shoulders with rocks, urging their men to do the same, when it was necessary to build a fortress near Antioch and a bridge on the Euphrates, respectively.[112] It is not a coincidence, of course, that these anecdotes appear only in promotional narratives; they probably derived from lost sources, and their historicity is highly doubtful.[113] But the image that these sources conveyed does not seem to be restricted to Byzantium, as it also occurs in Arab promotional narratives. The poet and close associate of Sayf al-Dawla, al-Mutanabbī, reports that during the reconstruction of the walls of al-Hadath in 954, Sayf assisted his men in the work by building with his own hands.[114] The appearance of relevant material in the Arab sources raises questions of whether these ideals were universal or whether the Arabs had access to Byzantine material.

Although most aspects of generalship found in historical narratives are regarded as moral and symbolic, there are some, such as confidentiality and hand-to-hand combat, which may shed some light on whether they were indeed followed in the battlefield. These are helpful not only because they are more practical but also due to the existence of parallel accounts which speak in favour of their implementation. To begin with, the *ST* advises the general to keep his judgement for a course of action secret until it is time to apply it. It also dismisses those 'who share secret plans (. . .) with all the soldiers' as 'senseless and with imperfect intelligence', explaining that if plans are widely disseminated, they end up being reported to the enemy by deserters.[115] It is no surprise that this trait appears in Byzantine historical narratives: Psellos reports this advice almost verbatim as given by Bardas Skleros to Basil II during their meeting in c. 991, quoting Skleros as saying, 'share with few your most intimate plans'.[116] While this passage could have been included for encomiastic purposes, such as to justify Skleros' career or facilitate the rehabilitation of the family,[117] it also appears in an Arab source which appears to be a military dispatch written in 903 by Muḥammad Sulaymān for his victory over the Qarmatians. We read that as soon as the commander learned about the exact location of the enemy, he 'kept this information concealed from the officers and everybody else and did not reveal it'.[118] Although we have seen that military dispatches have their weaknesses as sources, it is safe to assume that the mere need to refer to his concealment of information may suggest that this was indeed an accurate reconstruction of the way that commanders were expected to act.[119]

As far as hand-to-hand combat is concerned, the approach of the military manuals was that 'the general should fight prudently rather than daringly, or he should altogether refrain from coming to blows with the enemy'. The *ST* explains that the same stance was expected to be adopted during a raid, possibly in an attempt to make the passage more relevant to contemporary practice, since the yearly raids were a paramount aspect of Arab–Byzantine warfare.[120] *MS* and *LT* record the same mentality, since they instruct the general to draw up in the middle of the second line and discourage him from fighting with his own contingent.[121] This is reasonable enough, as the death of a general in a medieval context could prove disastrous for the fate of battle. In the same line of reasoning Nikephoros II proposes in the *PM* to aim the charge of the *kataphraktoi* directly against the enemy general, which is likely to make the enemy flee.[122]

The extent to which Byzantine generals refrained from hand-to-hand combat is difficult to assess; depending on who did the fighting, most Byzantine histories paint a contradictory picture. If historical narratives draw on promotional sources and have a positive disposition towards a general, the subject is, on the one hand, considered brave and is pompously described as taking part in heroic deeds or single combat, whereas a few lines later, the same subject is portrayed as a careful and prudent general who followed the advice of manuals. This contradiction served to paint the picture of the perfect soldier as the military aristocracy perceived him. The latter was influenced by the dual imagery of Achilles and Ulysses in the *Iliad* and *Odyssey* or by the brave deeds of Roman warriors, as well as by the single combat and cunning warfare of the Old Testament.[123]

Some characteristic examples of glorified narratives can be found in the histories of Leo the Deacon and John Skylitzes. When reference is made to central figures of their work, like Bardas Skleros, Nikephoros II Phokas and John I Tzimiskes, we find detailed descriptions of hand-to-hand fighting and we read that 'he always used to fight (. . .) in the van of the army, ready to meet any danger that came his way, and ward it off valiantly' or that 'he was not afraid of attacking single-handed an entire enemy contingent and (. . .) would return again (. . .) to his own close formation'.[124] Even these narratives, however, make sure to compromise bravery and heroic fighting by contradicting the image they had already set for certain protagonists with acts of prudent generalship.

Less famous figures did not seem to have enjoyed a given tolerance or admiration for behaving more daringly. A famous tenth-century example is the commander of Adrianople, Leo. During the siege of 921 by the Bulgarians, Theophanes Continuatus reports that, due to his daring behaviour, Leo was called 'Moroleon' (i.e. foolish Leo). Theophanes comments against this nickname and argues that Leo should instead be called 'Thymoleon' (i.e. brave Leo).[125] Other historians do not seem to share his view though; Symeon Magister refers to Leo as Moroleon, although he does record that he accomplished worthy deeds against the Bulgarians, and John Skylitzes describes Leo as very rash and calls him 'Moroleon'.[126]

This dissention demonstrates that the heroic image of Byzantine generals most probably served encomiastic purposes. Its aim was to present prominent members of the military aristocracy in an ideal light, rather than to give a realistic representation of the way in which generals acted on the battlefield. Except for promotional histories, the majority of Byzantine chronicles narrating events up to the middle of the tenth century do not provide much evidence of generals being involved in direct fighting. The testimony of Arab sources seems to agree in part with this less heroic image, because whenever there is a description of an expected battle, the Byzantine general is not involved into the fighting.[127] For instance, in the battle between John Kourkouas and Sayf al-Dawla in 938, Sayf managed to emerge victorious against the Byzantines but had to penetrate deep into the Byzantine troops before he could reach Kourkouas.[128]

To sum up, the profile of a general is something which appears to have largely been influenced by promotional literature or biases in the historical narratives. Most of the qualities of the ideal general which appear both in Byzantine histories and in military manuals are moral, and they serve as a means to idealize figures. In turn, the lack of morality and prudency is employed to disapprove of certain generals and to justify defeats most probably because the lack of these qualities can also be regarded as a sin in the Christian and Muslim religion.[129] Some aspects of generalship, such as avoiding battle and not sharing information, seem to have had some practicality as they feature in both Arab and Byzantine sources. In any case, there is little doubt that the *ST* provides an updated version of how the ideal general should be, a version which seems to be in accordance with the concept of Christian morality and the ethos of the military aristocracy in the tenth century and beyond.[130]

Practical information 2: punishments and rewards

The author of the *ST* dedicated two chapters to the discussion of military punishments. His material draws heavily on *MS*, while similar passages are also found in *LT*.[131] In the version of the *ST* we read that

> the general must moderate the offences of the rank and file and he should not aim his punishments at the masses so that common discontent may not unite them in revolt. At any rate, he should only inflict sentence and punishments on those who were the ringleaders of indiscipline.[132]

The importance of this measure is well highlighted in the historical narratives. John Skylitzes reports that Leo V (813–820) would punish everyone with extreme severity in order to instil fear, and thus 'earned the hatred of all his subjects', while Michael Attaleiates records how Romanos IV's excessive punishment to a soldier made a very bad impression.[133] Attaleiates' statement for an emperor whom he sympathizes with and Skylitzes' comments, even if exaggerated to present the iconophile Leo V as a monster, seem to denote that the advice of the *ST* reflected mainstream values.

The authors of Byzantine historical narratives suggest that this mentality was implemented on campaigns. Leo the Deacon and Michael Psellos delineate Nikephoros II and Basil II playing it by the book and inflicting severe sentences only on instigators. During a march to Tarsus (965), Nikephoros II, we learn, ordered a Byzantine officer to punish a soldier who dropped his shield by cutting his nose and parading him through the camp. When the officer disobeyed his orders, Nikephoros II ordered the same punishment to be applied to the officer in order to discourage similar behaviour by others.[134] For Basil II we know that he 'promptly discharged' and 'punished like common criminals' certain individuals who rashly charged the enemy, disobeying his order to hold the line.[135] Given that the histories of Leo the Deacon and Michael Psellos include material which has a positive disposition towards these two emperors, it is best to treat their accounts with caution.

Perhaps more credible are the accounts of earlier chronicles which do not appear to be so influenced by the impact of promotional literature of the military aristocracy and which present events with what seems to be a neutral tone. These chronicles describe that Michael II (820–829) managed to deal with the revolt of Thomas the Slav by giving amnesty to the masses that supported the rebel. As a result, the rank and file apprehended Thomas, who was severely punished by amputation of all his limbs. The amnesty, along with the punishment of the leading figure alone, led the rebel garrison of Bizye to act accordingly, seizing Thomas' son Anastasios and surrendering him to Michael.[136] John Kourkouas undertook a similar course of action during the revolts against Romanos I. All the chronicles agree that Kourkouas managed to suppress the revolt of Bardas Boilas and Adrian Chaldos, and after his success 'he only blinded the most important of the men he arrested, and confiscated their property, but the poor and insignificant he let go scot-free'.[137] Kourkouas' actions facilitated in achieving unity and ensuring that no new ringleaders would take over.

The lenient treatment of the rank and file does not only occur in these historical accounts but is also supported by Arab sources. For example, the *ST* makes specific reference to the fact that 'if the enemy is already approaching and a pitched battle is expected, it is appropriate for the general at that time to be very remiss about punishing the soldiers who commit offences, (. . .) mitigating, as far as possible, those who are under suspicion, as well as having a disposition towards leniency'.[138] This way of thinking was applied on the battlefield by Emperor Theophilos. According to the testimony of Byzantine sources, the emperor was in battle against the Arabs when the majority of his army was routed and thus abandoned him in the field with only a small unit of men. When Theophilos managed to escape, he 'limited himself to scolding the army which had abandoned him, he did nothing else unpleasant'.[139] This narrative agrees with Arab sources which also stress the lenient approach of Theophilos towards his soldiers in the midst of war. Al-Tabarī reports that prior to the siege of Amorion, part of the Byzantine army did not manage to unite with the emperor when a battle took place. Theophilos learned that this force was routed by the Arabs and that the Byzantine commander in charge had abandoned his troops. Afterwards Theophilos arrested him and had him decapitated, but since the campaign was far from over, he ordered the Byzantine garrisons of surrounding fortresses to punish the soldiers only with lashing and then to point to them the location of the new rally point.[140]

While some sort of punishment was usually reserved for offences and indiscipline, those who fought bravely and fulfilled their duty received some kind of reward; the most distinguished were rewarded both with material and honorary rewards. Slightly revising the material of *LT*, the *ST* states that 'it is fit for the general to give benefits to those who acted bravely during the battle, promoting some to a higher rank, or giving money, or land grants to others' and that distinguished men should be rewarded 'with banquets and breakfasts, some [given] by the general himself and others by their commanding officers'.[141]

Material rewards feature quite frequently in Byzantine historical narratives. The issue here is that most of them appear in histories which include promotional material and, to make matter worse, in a rather fixed way, leaving one to wonder whether they are mere clichéd references. For example, John Skylitzes and Leo the Deacon inform us that Theophilos 'awarded gifts and various honours to Theophobos',[142] that Nikephoros II Phokas 'handed out donatives to the host as was fitting' after the siege of Mopsuestia (965),[143] and that John I Tzimiskes, after the end of his campaign in 971, offered gifts to his victorious army, 'as was fitting'.[144]

Although this evidence cannot be conclusive, legal sources appear more credible because their purpose was practical, that is, to enforce a policy, rather than to describe events favouring political figures or inveighing against them. Information found in legal texts substantiates the advice of the manuals. A novel issued by Nikephoros II Phokas reports that abandoned lands were given to soldiers who performed valiantly in battle, demonstrating that the distribution of land grants to distinguished soldiers was a measure that was indeed followed.[145]

As far as honorary rewards are concerned, estimating their practicality falls to the same problem of evaluating the narrative of promotional sources. Historical narratives, like that of Leo the Deacon and John Skylitzes, describe emperors and generals as granting distinguished soldiers promotions and invitations to banquets. Their references, however, are very suspicious, not merely because they appear in conjunction with eulogized figures, but also because they are written in a uniform manner and, aside from that, in a wording similar to that of the manuals. For instance, we read that after the great victory of Leo Phokas against Sayf al-Dawla (960), Romanos II gave promotions and distinctions to those who fought bravely under Leo Phokas[146] and that after the battle of Dorostolon (971), John I Tzimiskes rewarded and entertained his soldiers 'with sumptuous banquets', as well as granted them promotions which 'made them even more zealous for battle'.[147] Perhaps a more exaggerated attempt to depict an individual as an ideal commander comes from Michael Psellos, who records that the soldiers of Bardas Skleros were very loyal to him because he always ate and drank with them.[148]

Even though it is correct to raise caution for such passages, dismissing the advice of manuals as impractical would be a hyperbole. There seems to be little doubt that such measures were more or less established as relevant references, and different kinds of sources seem to agree with this approach. To begin with, although military banquets are found in historical narratives which include promotional material, some passages mention them out of the context of praise and military activities and unrelated to victories and triumphs. For example, the *Vita Basilii* casually and neutrally reports that the *domestic* of the *scholai* Antigonos invited prominent military and political figures to his banquet, while all chronicles agree that during the conflict with the Bulgarians (922), Romanos I gathered the various commanders of the army in a banquet 'urging them valiantly to go forth against the foe'.[149] Since this banquet inspired a surprise attack which was eventually a failure, it would be inapt to argue that the purpose of this was passage was to portray Romanos I as an ideal general and energetic ruler. The most characteristic case demonstrating that military banquets were commonplace is the fact that they even appear in anecdotes which belonged to legend. We read, for instance, that a gypsy predicted the accession of Michael II to the throne, which led his military commander to invite him to a banquet 'to the exclusion of all others, even of those who were of superior birth and rank'.[150]

The same applies for promotions as well; the fact that such rewards were ordinary is underlined by the fact that they are also mentioned in military orations. A military speech written most likely in 950 by Constantine VII[151] informs us that 'the commanders of the *tagmata* and the other units who fight courageously will be rewarded in proportion to their deeds, some to become *tourmarchs*, others *kleisourarchs* or *topoteretai*', while even the common soldiers 'who display the traits of valour, will receive the due reward'.[152] As the purpose of this text was to be read in front of campaigning soldiers to boost their morale, it would have been out of place to refer to a measure that had no practical application whatsoever.

All these generous rewards concerned the lucky few who distinguished themselves in battle. For most soldiers, whose services were taken for granted and not regarded as exceptional, the most common reward was a share of the spoils. According to the *ST*:

> The distribution of spoils should be made equally among those who engage in fighting: likewise, among those who are on guard behind them or among those who guard the baggage train. They should be given both to the lower [ranks] and to the higher, because this is the law for the whole army. The general should not receive a portion from our majesty's grant more than the old tenth or the present sixth.[153]

The exact extent to which this advice reflected a tenth-century reality is difficult to assess. Historical narratives are unreliable not so much because they portray generals as following the advice of manuals but mainly because they give generic information without referring to specific figures. We read, for instance, that after his victory, Leo Phokas 'distributed most of it [the booty] to the army' or that after the capture of Chandax (961), Nikephoros II Phokas 'handed over everything to the soldiers as plunder'. The rest was 'displayed at the triumph (. . .). Then he deposited the wealth of the barbarians in the public treasury'.[154]

Legal sources, on the other hand, are once again more valuable. The author of the *ST* presented his information as deriving from some kind of legislation using the phrase 'στρατῷ νόμος'. Indeed, the passage is based on the *Procheiros Nomos* (870–879 or 907), which seems to demonstrate that the measure of spoil distribution was legally enforced.[155] It is, however, unclear whether the sixth was actually the standard figure to be reserved for the public funds in the period between 920 and 944, for other texts, such as *LT*, mention the fifth.[156] Equally unclear is the contemporary value of the rest of the passage. The author of the *ST* copied from the *Procheiros Nomos* that 'the increase in salary should suffice', but he added 'and the plentiful grants which are given to them on each [occasion] by our God-crowned majesty'.[157] The only available hint to determine whether there was actually some increase in the salaries of officers comes by comparing the information provided by Ibn Khurdādhbih, the *DC* and Liudprand of Cremona concerning the payment of officers. Judging by their accounts, it seems that the payments had been reduced at some point in the reign of Leo VI and then restored back to normal.[158] Be that as it may, Byzantine chronicles may supplement this view since they record a number of generous expenditures and donatives made by Romanos I and Constantine VII.[159]

In conclusion, the advice of the *ST* concerning punishments and rewards seems to fit the challenges and context of tenth-century Byzantine warfare. While it is impossible to argue that it was always followed verbatim, there is evidence which points towards its application. Lenient punishments are recorded in both Byzantine and Arab sources. Rewards and distribution of booty are not only validated by historical narratives, which one might argue were influenced from biased accounts, but also from other types of sources, most importantly legal ones, whose purpose was mainly practical, or at least intended to be. However, certain details such as the amount of spoils that went to the public treasury or whether the payments of officers were indeed higher before the middle of the tenth century seem to be less clear.

Practical information 3: gathering intelligence

In a world where conflict was endemic and guerrilla tactics and raids played an important role, sometimes taking place even three times per year, warfare was dominated by manoeuvres, ambushes and ruses. In this context the gathering of information could prove both challenging and vital for a successful campaign.[160]

The *ST* advises the general to obtain information about his surroundings and the routes that the enemy is intending to take before and after the battle.[161] A good example of how negligence in this matter could prove fatal is given by Arab sources which narrate the events that took place in 953. After Sayf al-Dawla had finished ravaging the area of Melitene, he tried to retreat following the *kleisourai* southeast of the city. Constantine Phokas, the son of the *domestic* of the *scholai* Bardas Phokas, managed to follow Sayf closely and blocked his way by capturing the mountain passes with his infantry. As a result, Sayf was compelled to retreat and find another way to cross. The Byzantines took advantage of his absence and ravaged the region of northern Syria and Antioch. However, Sayf was well informed about this turn of events, and by forcing his way across the Euphrates, he surprised the Byzantines by returning and attacking them. The latter, who were not informed about Sayf's whereabouts, were taken by surprise with his manoeuvre. This resulted in a complete defeat; the Byzantine army suffered heavy losses, among them, the *patrikios* Leo, son of Maleinos, while other *patrikioi* and Constantine Phokas were taken captives. The *domestic* of the *scholai* himself was injured in the face, and all the Arab prisoners and booty were recovered by Sayf.[162] However, the tables turned in 962, as Sayf was the one who failed to discover which route Nikephoros II Phokas followed, which resulted in a sudden engagement outside Aleppo, where Sayf was utterly defeated.[163]

To avoid such a predicament, a way to gather intelligence was required, the most obvious one being espionage. The *ST* states that the general should not act before he is well informed about the matters that concern the enemy. It continues to report that this can be achieved

> by always sending spies throughout their camp and by placing them in ambuscades, as well as by reconnoitring the nearby places for lying in wait, in case the enemy might be hiding in them, and by taking care of everything that regards the safety of his army.[164]

Similar information is, of course, recorded in earlier manuals such as *MS* and *LT*.[165]

In order to determine whether the Byzantines took this measure into consideration, one has to look for parallels in other sources. The history of Leo the Deacon records that this advice was followed almost verbatim. We read that Bardas Skleros 'summoned John Alakas and sent him out as a scout to observe the Scythians, estimate the size of the host, and see where they have camped and what they were doing' or that he 'immediately sent spies disguised as beggars to the camp of Bardas Phokas',[166] while we are informed that John Tzimiskes ordered his generals to send spies to the enemy camp and then to inform him about their plans.[167] Last but not least, we learn that Nikephoros II Phokas sent Nikephoros Pastilas to scout the surroundings immediately after he disembarked on Crete and that Leo Phokas spied on the camp of the Hungarians to find out their numbers.[168] On a closer look it seems as if Leo the Deacon wanted to communicate that all the generals who are positively portrayed in his narrative (the Phokas family, John Tzimiskes, Bardas Skleros) acted in accordance with the manuals.[169] Though Leo the Deacon could have drawn on military manuals or lost promotional sources to idealize the Phokades, other historical narratives may have done so through imitation. A characteristic example is Theophanes Continuatus' account of the conquest of Crete by Nikephoros II Phokas. Theophanes agrees with Leo the Deacon that Nikephoros II employed spies in the first stage of the operations, but his narrative had been most probably modelled on Procopius' *Vandal War*, in which Belisarius is also recorded to have used spies early in the conquest.[170]

To remedy the weaknesses of these sources, one has to turn to less questionable accounts. To begin with, Byzantine chronicles casually record that the Byzantines spied on the movements of the Bulgarians during the reign of Romanos I, and their testimony is corroborated by Arab sources which mention that Byzantine spies were caught prior to the poorly documented Byzantine expedition against Egypt in 926.[171] More significant is the evidence found in administrative documents. In the lists of the *DC*, which seem to have been compiled for utilitarian purposes, we read that the *archon* of Cyprus was responsible for sending spies to Tarsus and Syria prior to the Cretan expedition of 911.[172] Consequently, despite the biases and weaknesses of certain historical narratives, espionage was most probably regularly employed in the tenth century.[173]

Information on how these spies were chosen is also found in the manuals. Relevant passages are found in earlier treatises such as in *MS*, *LT* and the *PS*, but most of these are incorporated in the *ST* which states that spies 'must be prudent and bold, know the customs and the language of the enemy and also have a precise knowledge of the roads and topography'.[174] The knowledge of the Arabic language appears to have been an indispensable skill, allowing spies to mingle unmolested with the enemy. It seems that such men were indeed available to the Byzantines, if needed, and this appears in both Byzantine and Arab sources.[175] For instance, Leo the Deacon states that John Tzimiskes ordered bilingual spies to be sent to the enemy, and al-Ṭabarī records that during the siege of Amorion (838), the Byzantines tried to send a letter to the emperor, giving it to a man who could speak both Arabic and Greek, hoping that he could pass through the enemy lines without trouble.[176]

As far as the strategy of spies is concerned, the *ST* instructs that 'when we want to send them to conduct espionage, we take each one privately, and suitably instruct them one by one about those things which they should know, so that they may communicate with one another under the pretence of buying or selling when they are in hostile territory, and thus reveal what is happening'.[177] The information that spies acted under the pretext of trade or approached each other in the markets is also referred to in more detail in the *DV*.[178] While there is no similar reference in Byzantine historical narratives, Arab sources seem to agree that such methods were indeed employed by Byzantine spies. Ibn Ḥauqal, who was a contemporary to the events he describes, reports that during Romanos I's reign the Byzantines sent some merchants by ship. The merchants started to gather information about the regions and the state of affairs in the frontier, and under the pretext of conducting business, they met with a number of important Arab figures and then reported back to the Byzantines.[179]

Although spies were a vital source of information, they were certainly not the only one. Intelligence could also be gathered by other means, one of which was through defectors or traitors. The *ST* warns the general that he must be suspicious of defectors 'even if some (. . .) come (. . .) proposing an attack or promising to lead the way through roads unknown to most men, (. . .) or to unexpectedly fall upon the enemy', while it also notes that defectors should not be readily trusted because they might give false information which had been planned beforehand.[180] The need to practise caution with the words of Arab defectors seems to have been a contemporary practical issue which is also confirmed by Arab sources. For instance, we know that in 932 the Byzantines missed a great opportunity to re-capture Melitene since they intended to launch an expedition, only to be discouraged by an Arab defector called Bunnay b. Nafīs. The circumstances were very favourable for the Byzantines: the cities of the Arab frontier had revolted against the caliph in protest against his neglect of their protection; there was a significant disruption of foodstuff and supplies, not to mention that a leading figure of the Abbasids, Mu'nis, had started a civil war in his attempt to install a new caliph on the throne. It seems that Bunnay b. Nafīs, who was also an associate of Mu'nis, was asked by the latter to discourage a Byzantine attack so that Mu'nis could focus on achieving his internal-policy goals.[181]

Other contacts with defectors were more fruitful for the Byzantines. John Skylitzes reports that Niketas Chalkoutzes, who probably accompanied Sayf al-Dawla to his expedition in 950 as a hostage, kept Leo Phokas constantly informed about the routes and plans of the Arabs and so contributed to the Byzantine victory that year.[182] However, neither Leo the Deacon nor Theophanes Continuatus mentions the contribution of Chalkoutzes.[183] Similarly, al-Tabarī records that Theophilos attacked and captured Zapetra in 838 after he was urged to intervene by al-Bābek. Al-Bābek welcomed a Byzantine invasion since he had rebelled against the caliph earlier in the year, and being hard-pressed at that time, he looked forward to any distraction of the caliph's' troops.[184]

In addition to defectors, captives and prisoners of war were used for gathering intelligence. In fact, the *ST* suggests that prisoners of war 'who have been captured by assaults are to be more easily believed, since it appears that they give unprepared responses to our questioning and that they do not lie as much'.[185] Byzantine narratives seem to confirm that captives were a very effective means of acquiring information. Leo the Deacon states that during the Cretan expedition, Nikephoros II Phokas was informed by war prisoners that the Arabs were gathering forces to attack him unexpectedly, hoping to take him by surprise.[186]

Arab sources agree that interrogating prisoners of war to acquire intelligence was a standard and indispensable practice. In 838, the Arabs wanted to learn about the whereabouts of the Byzantine army, and thus they decided to take captives in order to interrogate them. The captives informed the Arabs of the location of the Byzantine army and of the fact that the nearby fortresses were aware of their plans and position.[187] During the same campaign, a serious lack of provisions led the Arabs to kill Byzantine captives. According to al-Tabarī, one of them asked to be spared, proposing in return to reveal the location of Byzantine fugitives who were carrying supplies with them. The Arab general, al-Mutasim, agreed to follow the way proposed by the old man and to grant him his freedom should his words prove truthful. This account is similar to the advice of the *ST* on how to act on such cases. More specifically, the *ST* states that if the traitors propose to lead the army through unknown passages:

> The general must keep them under observation after enchaining them close to him. He must safeguard himself with the most horrible oaths that if they tell the truth and do everything for the safety and victory of his army, he will release them from their bonds and that he will provide them with worthy gifts, but if they lie and prove guilty of desiring to put our army into the hands of the enemy, they will be immediately butchered with swords, limb by limb, by those who are guarding them.[188]

Al-Tabarī goes on to report that the Arabs started to suspect that the elderly Byzantine prisoner was leading them in circles, at which point they thought of killing him. The old man, with the help of another two captives, finally managed to find the Byzantine fugitives he promised, and he and the other two captives were set free.[189] A similar account is given by Ibn al-Athīr, who states that during the Byzantine–Arab conflicts in Sicily (859) the Arabs took a number of prisoners, one of which was a Byzantine official. The Arab commander al-Abbās ordered that all the prisoners be slain, but the official asked for his life to be spared and in return promised to give information on how to capture Castrogiovanni. The Arab commander accepted and ordered a detachment to follow the lead of the prisoner, who during the night showed them a secret doorway close to the sewers.[190]

A final source of information is refugees. Based on Onasander and *LT*, the *ST* warns that refugees might inform the enemy and recommends that

> if the general is setting out to have a city delivered by betrayal at a specific time, he should first capture and deal with those whom he encounters on his way, so that none of them may inform those who are inside about the assault by running on ahead, and so that our army's assault might be completely unexpected to them.[191]

In the context of yearly raids, which frequently aimed at storming minor and major fortresses, this advice was more than relevant. Arab and Byzantine sources suggest that generals took this matter into consideration whilst on campaign. As reported by Yahya of Antioch, the *domestic* of the *scholai* Bardas Phokas besieged al-Hadath (955) and blocked all the surrounding paths in order to prevent the locals from informing Sayf al-Dawla of the siege. Although Bardas was successful in that, it was eventually the absence of all news that made Sayf suspicious and prompted him to march towards al-Hadath, which he successfully relieved.[192] According to John Kaminiates, the intention of the Arabs to sack Thessaloniki (904) was reported to the Byzantines by refugees.[193]

Since intelligence could make a significant difference in the outcome of war, the gathering of information could not go unchallenged. The enemy would sometimes employ counter-intelligence, and apart from misinformation employed by fake deserters, false intelligence also involved the spread of rumours, false news and propaganda. In the context of endemic warfare, the author of the *ST* felt it was necessary to reproduce *MS*'s warning that

> the unpleasant rumours, which are spread by the enemy or even by us, must be closely scrutinized and not ignored, because these rumours are often true. Entire armies suddenly fall into the greatest dangers, when the general is neglectful towards them and he does not take the proper precautions.[194]

Byzantine and Arab sources affirm that false rumours were regularly circulated and could sometimes prove fatal. Constantine VII makes explicit reference to this practice in his military oration:

> In truth the Hamdanid has no power. Do not believe in his skills and wiles, he is afraid, he is devious and without a reliable force (. . .), he is trying to put fear in your minds with ruses and deceptions. One moment he proclaims that another force is on its way to him and that allies have been despatched from elsewhere, or that from another quarter a vast sum of money has been sent to him, while at other times he has exaggerated rumours spread about for the consternation of his listeners.[195]

Despite the reassuring mood of Constantine VII, the act of misinformation is portrayed as a realistic threat. Byzantine chronicles also suggest that this danger was beyond mere rhetoric, as they record that during the reign of Michael III (842–867), Theoktistos was successful in fighting the Arabs of Crete until they managed to persuade him that supposedly there had been a coup and a new emperor was reigning in Constantinople. Consequently, Theoktistos left Crete to return to Constantinople, leaving behind a portion of his army, which being left without sufficient leadership, was destroyed by the Arabs.[196] According to the testimony of Arab sources, around 955 some Byzantine spies seem to have been fed false information which resulted in the demoralization of their army,[197] while Sayf al-Dawla most probably spread false intelligence to al-Ikhshīd (944) in order to make him appear as the aggressor and therefore as responsible for the breaking of their political alliance.[198]

To sum up, although most of the passages of the *ST* which discuss the gathering of intelligence have little or nothing innovative, they remained more than relevant in the context of tenth-century warfare. Historical sources record many similarities with the manuals. While some narratives are problematic due to biases and *mimesis*, their testimony is corroborated by Byzantine military orations and administrative documents, as well as by Arab historical narratives, which seem to underline the practicality and application of such practices.

Practical information 4: camps and night attacks

Almost all Byzantine military manuals dedicate a passage or two on how to pitch a camp securely while on campaign. The *ST* is no exception to the rule, drawing upon earlier manuals its author reports that for the security of the camp:

> The general should surround the remaining site with an artificial trench (. . .). The gathered soil from the trench should not be thrown outside but inside, that is to say towards the side of the encampment. Whenever the ground is harsh and cannot easily be dug, he should secure the camp in every direction either with a wall made of bricks, or stones, or tree trunks, or with fences, or with a large number of wagons, above all due to the enemy surprise attacks and especially ones at night.[199]

The historical narratives which describe such practises in detail are that of Leo the Deacon and John Skylitzes. The reader of these histories gets the impression that securing the camp was more or less a standard procedure in the Byzantine army, practised by almost every general. When Leo the Deacon has a positive disposition towards a general, he presents him as applying the advice of the manuals word for word. Therefore, it comes to no surprise that John Tzimiskes secured his camp with a ditch and used the gathered soil as a wall, in which he constructed a palisade with spears,[200] neither that Nikephoros II Phokas 'pitched a camp and surrounded it with a palisade', nor that 'he fortified it strongly all around with a stockade and a ditch' during his Tarsian and Cretan campaigns.[201] Leo the Deacon sometimes goes as far as to describe that the Byzantines not only secured their camps but also picked ideal locations for them. Conforming to the recommendations of the *ST* by which the most suitable locations are those which 'have drinking water nearby' or 'where there will be a canyon or an inaccessible river or some other rough ground at the rear of those who encamp',[202] Leo the Deacon reports that John Tzimiskes rested his soldiers 'on a secure hill that had a river flowing past on both sides' the night before the siege of Preslav and that Basil II 'pitched camp in a thicket and allowed the army to rest'.[203] The fact that the narrative of Leo the Deacon is evidently similar to that of the manuals leaves the historian to wonder whether Leo the Deacon eulogized the actions of generals to show that they followed the Roman ideal.[204]

Alternative historical narratives cannot be taken at face value either because they present other complexities. Take the case of Nikephoros II Phokas and the Byzantine camp at Chandax, for instance. John Skylitzes agrees that Nikephoros II had 'set up a strong palisade surrounded by a deep ditch fortified with stakes and staves'.[205] Given that Skylitzes' coverage of Nikephoros II Phokas' reign is notably more moderated than that of Leo the Deacon, one could argue that indeed that was how Nikephoros II acted. On the other hand, promotional sources are also discernible in Skylitzes' narrative, which makes it probable that the similarity of information is nothing more than a dependence on some common source.[206] To make matters worse, John Skylitzes often covers military events with certain haste and standardized phrases, which makes his account less credible. Almost every time a Byzantine general reaches his destination, Skyliztes uses the phrase 'στρατόπεδον πήξας' in his narrative, which would suggest that the camps were always secured and established.[207] Apart from John Skylitzes, Theophanes Continuatus also covers the events at Chandax. While his testimony agrees that Nikephoros II 'χάρακα καὶ τάφρον βαθεῖαν κατασκεύασεν', his account is most probably modelled on that of Procopius' *Vandal Wars*, in which Belisarius is presented to have done the same.[208]

In spite of the fact that these narratives do not provide conclusive evidence for the security of the camp, other sources support that such measures were considered a standard procedure both in the Byzantine and the Arab world. In a passage that does not seem to draw on promotional sources or to imitate others, Theophanes Continuatus informs us that the Tarsians (833) 'found out that (. . .) neither had he [Stypeiotes] <dug> a trench and <built> a rampart in front of the encampment, nor had he accomplished any other of those things which thoughtful and sensible generals prepare in advance'.[209] In the same spirit, Yahya of Antioch records that Basil Lekapenos campaigned against Sayf al-Dawla (958), constructing a ditch around his camp everywhere he went, and speaks of how other Byzantine generals secured their camp in the same way, explicitly stating that this was a standard procedure.[210] What is more, al-Tabarī not only describes an Arab camp being secured in the same manner as a Byzantine one but also records how an officer supervised the watches in order to prevent surprise attacks. The latter brings to mind the advice of the *ST* which instructed the general to post and supervise guards at a considerable distance from the camp in order to be able to apprehend spies or to react to night attacks promptly.[211]

To reinforce the security of the camp, the *ST* and other manuals instruct the general to place caltrops in the ditch.[212] Although caltrops do not usually appear in historical narratives, the *Alexiad* records that Alexios used caltrops in order to secure his position from the charge of the Norman heavy cavalry (1082).[213] If we take into consideration that Anna Komnene idealized Alexios, one may wonder whether his actions had been deliberately modified to match the advice of the manuals. The caution for this passage is, however, moderated to an extent by a Norman source, William of Apulia, who agrees with Anna that Alexios used caltrops to secure his camp from enemy cavalry. William's account might only agree with Anna because he might have used a common lost source, but the contents and style of this work can only be guessed at.[214] Whatever the case, the practical use of caltrops seems to be backed by a number of textual and archaeological sources which demonstrate the extent to which this practice was applied by the Romans up to the fifth century AD and by the Arabs in the seventh and ninth.[215]

Although the manuals seem to have included practical and common measures for the security of the camp, Byzantine generals did not always conform to their instructions. Historical narratives often record defeats which were inflicted by enemy surprise attacks and night attacks against unfortified and disorganized camps, as well as against troops which were on camp duties, such as gathering forage and supplies, which denotes a certain negligence towards posting guards and sentries.[216] The *ST* makes explicit reference to this danger, stating that:

> Above all, however, the general must always keep an eye with devotion on this: that neither he, nor any of the picked officers with the stronger *tagmata* under their command should dismount from their horses, before the camp is safely completed and the watches set (. . .). For the greatest misfortunes befall the army precisely during such times, namely when it is occupied with establishing the camp or with the release of the horses for grazing or when it is dismounted and almost unfit for battle.[217]

Most historical narratives seem to imply that the advice of the manuals reflected an attempt to tackle real practical issues. A number of sources seem to provide a credible testimony, casually recording relevant events with a neutral tone. We learn, for instance, that Justinian II was defeated by the Bulgarians during the collection of hay since his cavalry camped without guard (707),[218] or that a Byzantine detachment under the command of Monasteriotes was ambushed outside of Tarsus as it was gathering forage (965).[219] Arab sources also record similar night attacks against the Byzantines, one of which was undertaken by Sayf al-Dawla, who was successful in defeating the Byzantines and reclaiming all the captured prisoners (959).[220] While there is little doubt that such defeats occurred, we are mostly in the dark regarding the details of the operations. The enemy night attacks are often covered in a standardized and vague manner in Byzantine narratives – most of the time we are merely informed that they occurred unexpectedly and that many Romans were killed. A significant exception is the defeat of Stypeiotes by the Arabs (833). As per usual, Byzantine and Arab chronicles only speak of a night attack which destroyed the Byzantine army, but the *Vita Basilii* records how 'men and horses were thrown together in confusion and fell upon each other'.[221] This seems to recall the advice of the *ST* which states that, in an organized camp, the infantry tents should be set in the perimeter and then, after a suitable interval, the tents of the cavalry. Thus, the cavalry has more time to prepare for battle and the horses are not taken by surprise so as to cause confusion or to run amok in such situations.[222]

Even though some sources are reliable, others should not be too readily taken at face value, and their testimony should be approached with the agenda of their author in mind. A characteristic example is the narrative of Leo the Deacon, which is infused with sympathies. As noted earlier in the qualities of the ideal general, morality seems to play a central role in the history of Leo the Deacon.[223] It is possible, however, that Leo the Deacon not merely blamed defeats on the bad character of certain individuals but also created events out of thin air so as to make his point obvious and to eulogize others in juxtaposition. For example, Leo the Deacon is the only source to record that during the Cretan expedition Nikephoros Pastilas was ambushed and defeated by the Arabs, the explanation being that Nikephoros Pastilas failed to keep vigilant and to post sentries due to his indulgence in luxury.[224] Leo the Deacon continues his account with a made-up speech delivered by Nikephoros II Phokas to the army in which he urges his men to vigilance and abstinence. Thus, Nikephoros Pastilas emerges as an immoral man and a scapegoat, and Nikephoros II Phokas as an ideal general who can inspire and lead his men to victory.[225] John Skylitzes does not mention Nikephoros Pastilas at all, let alone this event, and Theophanes Continuatus only reports that the place where the Byzantines camped was abundant in trees and food. It could be that Leo the Deacon was inspired by such a reference to manipulate the events and to create the story about Pastilas, though Theophanes Continuatus' account is open to question since it imitates Procopius' *Vandal Wars*, in which a very similar reference is found.[226]

Another good example is the way Leo the Deacon transfers the blame for Basil II's defeat by the Bulgarians (986). We read that 'the army fell into indolence and sluggishness as a result of the incompetence of the commanders'. Thus, 'the Mysians ambushed them [the Byzantines] first, when they left the camp for forage and fodder'. Leo holds responsible for the defeat some commanders he does not mention and the negative character trait of laziness which befell the troops.[227] Once again, there is no alternative testimony which speaks of this specific event. Although Leo the Deacon was an eyewitness to this campaign and his account is considered more trustworthy than Skylitzes',[228] we can note that a similar pattern emerges here: the event is found only in Leo the Deacon: some obscure figure(s) is blamed for the defeat, and a negative character trait is involved. It is therefore best to approach his narrative meticulously and to wonder whether this event is also fabricated.

While Leo the Deacon portrays obscure figures as ignoring the advice of manuals, glorified figures are narrated in comparison as warding off dangers and taking precautions. In his account it is not only their generalship that it is perfect, but even the time is ideal. We read that a Russian sudden attack on the Byzantine camp happened as 'evening was drawing on', which is exactly the time that the manuals instruct such attacks are more likely to happen. It comes as no surprise that the Byzantines managed to counter this threat successfully since John Tzimiskes was in charge of the army, and thus, playing it by the book, he had kept the camp fortified and his men vigilant.[229]

Leo the Deacon, of course, is not the only author who is biased. Historical narratives usually accompany Byzantine defeats by night attacks with comments underlining the inexperience or negligence of the commander in charge, like the aforementioned cases of Gongyles, Krateros and Stypeiotes, whereas some generals originating from powerful families enjoyed a more favourable treatment. For instance, the *domestic* of the *scholai* Leo Phokas was fighting against the Bulgarians (917) only to be surprised and defeated by a night attack. On the one hand, Theophanes Continuatus and Symeon Magister report this defeat without any negative remarks against him, and on the other, John Skylitzes says nothing of this predicament, recording instead that Leo Phokas emerged victorious in his struggles against the Bulgarians.[230]

Leaving aside sympathies and biases regarding defeats by sudden attacks and night attacks, there were times when the tables were turned, and it was the Byzantines who took advantage of the enemy's negligence. The *ST* dedicates a chapter on the issue of mounting night attacks against camps. What is more interesting for now is that the author of *ST* updated the tradition and included an original passage in which he described the credentials under which a night attack was considered honourable. According to the *ST* 'night battles were invented for times of weakness or shortage in the army. For if the army is fit for fighting, it [is] insulting and totally unworthy to win in such a way'.[231]

It seems doubtful that the honourability of a night attack actually concerned a Byzantine general before he mounted one. Rather than being central to the military way of thinking, the honourable conduct of night attacks seems to have been merely a tool at the hands of Byzantine historians to push their agenda, to highlight the qualities of certain figures and to promote and glorify the values as well as deeds of the military aristocracy. For instance, Leo the Deacon, who was pro-Phokas, makes explicit reference to the fact that Leo Phokas mounted a night attack against the Magyars in 960–961 because they 'enjoyed vastly superior numbers of troops, whereas he was leading a small and ill-prepared band of soldiers'.[232] Similarly, Theophanes Continuatus preserves an anecdote, according to which, Theophobos proposed to Emperor Theophilos 'a night-attack on the enemy by the infantry, with the cavalry being brought in as and when needed'.[233] Theophilos chose not to follow this advice because he was convinced that the motive of Theophobos was supposedly to diminish the glory of the emperor by making him attack during the night. The whole event, however, has nothing to do with practical credentials of honourable conduct of night attacks. The alternative testimony of Joseph Genesios records Theophobos' proposal but makes no reference to honour and glory. He merely states that the advice was not followed because Theophilos was already growing suspicious of Theophobos.[234] The whole anecdote seems to have been a literary creation serving to pick sides in the rivalry that existed between two generals, Theophobos and Manuel. Thus, Theophobos emerges as unreliable, and Manuel as worthy and loyal.[235]

The fact that honourable conduct of night attacks had insignificant operational value is also demonstrated by the fact that historical narratives do not describe Byzantine night attacks with negative remarks. The example of Nikephoros II Phokas is characteristic here. Despite the exaggerations of the sources, it appears that Nikephoros II had an able expeditionary force at his disposal, perhaps no fewer than 35,000 men if we judge by the previous Cretan expeditions.[236] But in spite of his adequate force, Nikephoros II mounted a successful night attack against the Arabs of Crete in 960, which not only lacks any negative remark, but on the contrary, is described as a triumph by Leo the Deacon.[237] Other military manuals such as *LT* and the *PM* corroborate this attitude as they dissuade the commander to participate in a pitched battle, even if the advantage is clearly on his side. More specifically, we are informed that:

> It is very dangerous, as we have frequently said, for anyone to run the risk of a pitched battle, even when it seems perfectly clear that <our forces> far outnumber [the] enemy. The result of fortune is unseen.[238]
>
> If the enemy force far outnumbers our own both in cavalry and infantry, avoid a general engagement or close combats and strive to injure the enemy

with stratagems and ambushes. The time to seek general engagements with
the enemy is when, with the help of God, the enemy has fled once, twice, or
three times and are crippled and fearful, while on the other hand our host is
obviously confident and their thoughts of valour have been awakened. Avoid
not only an enemy force of superior strength but also one of equal strength,
until the might and power of God restore and fortify the oppressed hearts and
souls of our host and their resolve His mighty hand and power. (. . .) When
She [the Virgin Mary] secure Her people's victory for the third time, from that
moment on they need not flinch or recoil in fear.[239]

Since night attacks were a suitable way to acquire victory with minimal losses,
military manuals expounded on how they ought to be mounted. We learn that 'the
best time for a night-attack is two or at most three hours before dawn and when
the night is full of stars or the moon is full', as well as that

the tacticians divide the whole army into only three divisions (. . .). They set
two of them on both sides of the enemy camp, with many bugles, trumpets
and copper drums, because in this manner the approaching army gives the
impression to the enemy that it [is] many times larger.

The treatises also explain that the third division should advance against the enemy
camp directly and that the camp must be attacked from three sides and not encircled so
that the enemy has a route to escape and does not fight bravely due to desperation.[240]

Byzantine histories record the mounting of night attacks in detail only when they
are conducted by a general they favoured. These operations are described in such an
ideal light that they seem to have been copied word for word from the manuals. To
begin with, the generals proceed with the attack under the ideal conditions. Thus,
Leo the Deacon has Nikephoros II Phokas conduct his when there was a full moon,
while the *Vita Basilii* states that the *strategoi* of the Charsianon and Armeniakon
themata conducted their night attack against Chrysocheir at the right time, just before
dawn.[241] The procedure of the attack is also flawless; Chrysocheir's camp was only
assaulted by one part of the army, the rest taking place on the slopes around so that
they 'would let out terrifying clamours with deafening war whoops and trumpet
blasts (. . .), in order to make it appear that vast numbers were involved', and Nikephoros II Phokas divided his army into three contingents 'ordering the trumpets to
sound and the drums to roll' just before the assault.[242] The only aspect of Nikephoros
II's generalship that deviates from the advice of the *ST* is that Nikephoros II is
recorded to have completely surrounded the enemy camp, something which could
have served to highlight the effectiveness and gallantry of Nikephoros' troops.[243]

An alternative way of attacking the enemy camp also finds an analogy in Byzantine historical writing. The *ST* stresses that

if there is a river flowing between both camps, especially if it [is] impassable
to the cavalry, commanders have quickly obtained victory when they suddenly
appeared against the enemy, after a bridge was built there and the army safely
crossed over it.[244]

John Skylitzes reports that Nikephoros Ouranos employed the exact same strategy against the Bulgarians in 997. According to Skylitzes' account, Ouranos pitched his camp opposite the Bulgarians' with the river Sperchios between them. Since the river was overflowing and an approach seemed impossible, Samuel neglected his guard. When Ouranos found a suitable spot to cross the river, he managed to reach the enemy camp at night and to slay the Bulgarians, who were still sleeping.[245] The appearance of promotional sources in Skylitzes' work and the lack of alternative accounts leave us to wonder whether this is a genuine employment of this stratagem or merely a promotional tool to idealize Nikephoros Ouranos.

The testimony of sources which approaches the deeds of the military aristocracy in a more detached way could have offered an important alternative but end up only informing us that the Byzantines did mount night attacks from time to time. A characteristic example is the narrative of Theophanes Continuatus where Pothos Argyros is simply recorded to have attacked the Magyars during the night, to have massacred them, and to have reclaimed booty and prisoners (959).[246] This casual and undetailed mention of events casts no light on how night attacks were actually mounted.

To sum up, we are compelled to approach most sources with caution when they refer to the organization of camps and surprise attacks. Promotional accounts like that of *Vita Basilii*, Leo the Deacon and John Skylitzes argue that successful Byzantine commanders applied the advice of manuals word for word. While these are our most detailed sources, they seem to describe an ideal conduct of operations, and in the absence of other parallel accounts, we cannot take their evidence at face value. Despite the difficulties, there is evidence to support that the advice of the manuals was followed in action. We are on firmer ground when we possess alternative accounts from more neutral Byzantine and Arab narratives, or even archaeological findings. These sources suggest that securing the camp was supposed to be a standard practice, but at times commanders were neglectful in taking precautions or posting sentries, and, consequently, suffered similarly to what the manuals describe. In addition, it seems that the warnings of the manuals served a very practical purpose and were perhaps included in response to failures.

Practical information 5: defensive siege warfare

Sieges were a very important aspect of Byzantine warfare, especially during the tenth century. In this period, the Byzantines were involved in numerous sieges against fortified cities and fortresses such as Melitene, Marash and al-Hadath, but they were also called upon to defend their own strongholds against seasonal Arabic raids and counter-attacks. Therefore, siege warfare included both defensive and offensive practices, which are covered separately in the *ST*.[247]

As far as defensive siege warfare is concerned, the *ST* provides the general with detailed advice covering almost every stage of a siege. To begin with, the general is instructed to take some necessary precautions before the arrival of the enemy. Of primary importance are measures of logistical nature. The general was expected to stockpile food supplies in order to be able to withstand a lengthy siege. Supplies were most probably gathered from the surrounding countryside, which had the additional benefit of preventing the enemy from supporting himself from local sources whilst he was besieging. As the *ST* puts it 'so that the enemy, may not (. . .) feed on fruits from the trees, if they have a shortage of food'.[248] Second, the general was to pay attention to his water supply; we read that 'if there is not an abundance of water or reservoirs in the city, drinking water may be enclosed in some kind of container or cisterns because (. . .) water must be measured and secured as much as possible so that it may not be easily snatched away by the more powerful'.[249] In case supplies were not enough to support a lengthy siege, the commander of the city was to 'send the sick, elderly, and women and children, to a safe and fortified location in advance of the enemy assault'.[250]

Byzantine and Arab sources suggest that the manuals offered routine advice with regard to supplies. We know, for instance, that during the reign of Michael II, the Arab forces were faced with famine when they besieged Syracuse, because the Byzantines had already gathered all the supplies from the countryside to the city and that the lack of water could prove to be the sole reason to force a city to surrender, as was the case with Geraca, where the Byzantines were compelled to negotiate a treaty with the Arabs (951).[251] Likewise, we are informed that Thomas the Slav 'expelled all the people who were unfit for service' from Adrianople since there were not enough supplies for everyone during the siege.[252] The Byzantines 'removed the surplus families' from Constantinople (714) and Amorion (716) too because they were expecting a siege, with the Arabs doing the same in Aleppo in response to the campaign of Romanos III (1028–1034).[253]

What is also interesting here is that the advice of the manuals was practical to such an extent that it seemed to annul the law. The fact that adult male civilians were excluded from those expected to abandon the city implies that they were recruited to assist in the defence of their city. Technically, this was in contrast with imperial legislation which prohibited civilians to carry, buy or sell weapons. The relatively small manpower of the Byzantine army, however, and the continuous hostilities strongly distinguished theory from practice. *LT* clearly instructs the general to encourage the use and the possession of at least one bow for every household in order to aid in the defence of the region. Additionally, civilians defending cities, from the capital to the borders, are attested throughout the course of Byzantine history. A clear tenth-century example can be seen in the siege of Thessaloniki by the Arabs in 904, where the civil population joined the garrison in the defence of the city.[254]

After food and water had been secured, the *ST* advises the general to undertake another preliminary measure: to 'cut down the trees which are near to the city walls and remove every kind of obstacle, so that the enemy may not hide in them'.[255] The practical application of this measure seems less clear as we do not possess a plethora of similar information in historical narratives. An exception is the testimony of Leo the Deacon, which mentions this measure in a slightly different context. Leo the Deacon records that in 965 the area around Tarsus and the Byzantine camp 'was filled with flowers and all sorts of trees'. Consequently, in order to secure his camp, Nikephoros II ordered his army 'to clear-cut and mow down thoroughly the fields (. . .) so that (. . .) it would be impossible for any of the barbarians to set up an ambush in thickly grown areas'.[256] While the history of Leo the Deacon often exaggerates the military deeds of Nikephoros II, his testimony seems to be supplemented by Arab sources which describe the extension to which the countryside was destroyed and report that 50,000 trees were uprooted.[257]

Nevertheless, securing the perimeter was by no means enough to ensure the safety of the city, for the greatest threat frequently lurked inside rather than outside. The manuals warn of this danger and request the general to 'pay serious attention to suspicious people'.[258] Historical narratives underline the importance of this advice, outlining the motives of these suspicious people. Some were driven by kin and alliances, others by religion, political faction or even by greed. For example, in an attempt to avoid treason just before the siege of Dristra, Sviatoslav 'put the Bulgars he had captured alive (. . .) in iron fetters and other kinds of restraints for fear they might mutiny'.[259] Amorion was betrayed to the Arabs by Boiditzes (838), either because he was persuaded by gifts or by the fact that he was an ex-Muslim prisoner who was compelled to embrace Christianity.[260] In turn, a Christian inhabitant of Amid proposed to the Byzantines to build an underground that would allow them to bypass the walls and enter the city. Shortly before the Byzantines were close to the walls the treason was revealed and the Arabs killed the Christian traitor (951).[261] A final example is the betrayal of Antioch in 969. John Skylitzes reports that Bourtzes managed to corrupt an Arab from the garrison and with his help constructed ladders which were able to reach one of the western towers of the Antioch, while Leo the Deacon and Yahya of Antioch speak of turmoil and division among the personnel inside the city.[262]

Arab and Byzantine sources record that sometimes treason was an act of spies who had managed to infiltrate the city hours before the siege. A good example is the capture of an unknown fortress in Greece by Symeon (918). After gathering intelligence about the state of the gates, Symeon ordered five men to enter the city on the pretext of going for work, armed with axes. The men entered the city, overcame the gate-watch and opened the gates to the rest of the Bulgarian army.[263] A similar attempt was undertaken by the Byzantines against Melitene (928). Melias attempted to take the city by infiltrating 700 men into Melitene before the siege. They were to act as if they were looking for employment and when the Byzantines arrived, they were instructed to betray the city. However, the officials of Melitene suspected the plot and decided to kill all unknown men who had entered the city recently.[264]

After securing supplies, the perimeter and the loyalty of the population, the final preliminary measure was to deploy and distribute the garrison. The *ST* records that the garrison should be assigned to various spots on the wall but also encourages the general to keep a force next to him so as to reinforce the more hard-pressed sectors. The guards were to be rotated as a means to avoid fatigue and to eliminate chances of treason.[265] Historical narratives seem to suggest that the instructions of the manuals took into consideration very practical issues or even past mishaps. John Kaminiates informs us that the commander of Thessaloniki posted some men on the walls, while he kept his bodyguards as a reserve in order to assign them the most vulnerable posts on the wall (904), and al-Tabarī states that the commander of Amorion divided his forces into contingents and appointed to their officers the task of defending a specific number of towers.[266] The same author indicates that negligence in these matters was the fatal blow that led to the capture of Amorion. A Byzantine officer, we learn, was very hard-pressed since he happened to be responsible for the sector where part of the wall had been breached. After he managed to ward off the Arabs by fighting continuously, he asked to be reallocated to another sector, but the commander refused to relieve him, even though other contingents were fresh. Feeling exhausted and mistreated, the officer betrayed his sector to the Arabs and defected to them.[267]

After all preliminary measures were taken, all the besieged could do was to defend the city against enemy assaults and siege engines as efficiently as possible.[268] The manuals offer guidelines on how to counter the threat of the most regularly employed siege engines. To avoid having the walls and gates breached by trebuchets and battering rams, the *ST* recommends to 'hang heavy mats or newly stripped-off buffalo hides from the battlements on the outer side of the wall in order to cover them easily, or timbers attached together like a textile' and to place 'very thick pikes and sacks full of chaff or sand' to soften the blows.[269] Against mobile shelters, the tortoises, the treatises suggest casting fire on them or boiling pitch and lead, as well as to 'build beams which have very sharp iron points at their ends because when they are stuck into the tortoises, they easily overturn them'.[270] To prevent the enemy from reaching the ramparts, the *ST* explains to the general that 'the so-called *strepta*, (. . .) which mechanically shoot the liquid fire, (. . .) and the so-called hand-siphons (. . .) get the better of wooden towers brought towards the walls with rolling cylinders'.[271] Finally, against ladders the general is advised to use 'mill stones and heavy timbers tied with ropes to every battlement' as well as 'pitch, oil and fenugreek'.[272]

The testimony of Arab, Byzantine and Western historical narratives complements the information of the manuals. Al-Tabarī records how the Arabs tried to breach the walls of Amorion by bringing their stone throwers to attack the most vulnerable section of the walls.[273] Most probably, the standard type of this engine at that time was the traction trebuchet.[274] John Kaminiates vividly describes the impact of such engines during the siege of Thessaloniki:

> Others applied themselves to stone-throwing engines and sent giant hailstones of rock hurtling through the air. Death threatened us in many shapes, and since it was coming from all directions, it lent a further dimension of terror to the experience (. . .) whose relentless fire made it impossible for anyone to venture forth with impunity on to the wall.[275]

It seems that the suggested measures for keeping the walls safe against trebuchets and battering rams were more or less standard, as al-Tabarī and Anna Komnene describe the Byzantines and the Turks, respectively, to have repaired the walls with timbers as well as packsaddles or to have protected them with mattresses, leather hides and clothes which served to soften the blows of the bombardment.[276]

Accordingly, Albert of Aachen records that sacks filled with sand or chaff were employed by the garrison of Jerusalem against the against the Crusaders' rams in 1099.[277] An exception to the rule, however, is the use of pikes in this context, which although recorded in many manuals, remains, most likely, without parallel in the historical narratives.[278]

As far as tortoises are concerned, their use seems to have been common. These mobile devices were indispensable, since men who tried to approach the enemy walls were vulnerable to missile fire, which, being released from the ramparts above, came down to them with higher velocity. To counter this problem, the besiegers usually built mobile shelters which were used as covers as they were trying to approach the wall with a siege ram or to undermine the walls. We explicitly know from *DAI* that their lighter version, the *lesai*, was already used by the Arabs during the reign of Leo VI, and Arab sources confirm that tortoises were employed against Amorion.[279] The use of pointed beams to counter the threat of tortoises does not seem to have been very widespread. Our sole evidence for the employment of this measure seems to come from John Skylitzes who gives us an account quite faithful to that of the manuals. More specifically, we read that the commander of the city of Manzikert, Basil Apokapes, supplied the Byzantine garrison with large beams which were sharpened at one end (1054). The Turks constructed light types of tortoises and started to approach the walls. When the time was appropriate, the Byzantines threw them off the battlements and thus overturned the enemy tortoises.[280]

The second way to deal with tortoises – using flammable substances – is better documented. The use of pitch and lead was already known from antiquity, and it certainly remained in use well after the tenth century. It appears that sometimes a flammable mixture which resembled liquid fire was stored and thrown in pots against the enemy.[281] Byzantine, Arab and Western historical narratives record the use of similar means. John Kaminiates informs us that some of the defenders of Thessaloniki tried to counter the Arab assault with artificial fire by preparing earthenware vessels which included 'pitch, firebrands quicklime and other flammable substances'.[282] Al-Tabarī states that during the siege of Amorion the tortoises of the Arabs were burned by the Byzantines, while Albert of Aachen mentions the use of boiling pitch in the east during the First Crusade.[283] Most importantly, the use of pitch in warfare is also confirmed by the testimony of administrative documents, as the lists of the *DC* feature pitch among the provisions taken for the Cretan expedition of 949.[284]

Flammable mixtures were also employed against siege towers and ladders. Byzantine and Arab sources demonstrate how the Byzantines used liquid fire to gain an advantage during a siege. For instance, we are informed by Michael Attaleiates that Basil Apokapes attacked and burned an enemy siege engine by throwing a pot which contained liquid fire, while Ibn al-Athīr records that John Kourkouas employed devices which shot liquid fire during the siege of Dvin in 927. Ibn al-Athīr also makes reference to the impact of these devices, stating that 'their fire could cover twelve people, and was so violent and so adherent that no one could resist it'.[285]

Unfortunately, the use of mill stones and heavy timbers is not as clearly documented. The most problematic seem to be the Byzantine sources, the account of which is both biased and generic. For instance, the *Vita Basilii* reports that the defenders of Euripos repelled the Tarsian raiders by 'throwing stones down by hand', among other missiles, and Leo the Deacon states that in the siege of Chandax the Arabs threw 'enormous stones' from the walls.[286] Despite the similarities, our accounts are suspect on the grounds that their promotional character may have begotten the fabrication of events. The testimony of John Kaminiates is also ambiguous, even though he was supposedly an eyewitness to the events he describes. John Kaminiates reports that during the siege of Thessaloniki, a detachment of the Saracens took a ladder and started to scale the wall, using their shields to cover their heads. The Byzantines responded, and 'a volley of stones as thick as hail was unleashed against them'.[287] One cannot help but notice that John Kaminiates, instead of a detailed original account, provided us with a generalized report of what happened, employing in fact a literary *topos* ('thick as hail') in his description. Consequently, one finds no explicit reference to heavy timbers and mill stones, although the fact that the enemy shields proved inefficient might imply the stones were quite heavy. In spite of methodological issues, the practicality of this practice seems to be confirmed by Arab and Western sources as well. Al-Tabarī generally reports that the Arabs were afraid of the stones thrown down by the Byzantines, while Ralph of Caen states that the defenders of Latakia threw heavy stones from the towers against the Normans in 1101.[288]

As a last resort to ward off enemy siege engines, the manuals instruct the general to strike unexpectedly against siege devices through postern gates. The author of the *ST* specifically states that this should only be done 'if there is a very great need'. Clearly the author of the *ST* regarded sallies as a risky tactic and generally advised the defenders not to 'fight outside of the walls, even if they happen to be greater in numbers and braver'.[289] This approach seems to correspond to the testimony of Byzantine histories like that of Leo the Deacon and Michael Attaleiates. We read that during the siege of Dorostolon (971), the Rus made a sally to attack the Byzantine siege engines, as 'they were unable to withstand the whizzing missiles the latter [i.e. the Byzantines] hurled',[290] and that Basil Apokapes made a sally and burned an enemy siege engine, after he was unable to counter it in any other way.[291] Both sources offer a narrative as close to the manuals as one can get. In the absence of alternative accounts, it is best not take their information at face value due to the fact that these narratives could have been deliberately modelled to match the instructions of the manuals in an attempt to idealize the military aristocracy and its endeavours.

Our caution, however, for the practicality of this measure is moderated by more neutral accounts. As we have already seen, Byzantine chronicles appear divided on the issue of bold and rash practices. Thus, according to the opinion of some, Leo, the commander of Adrianople, deserved to be called 'stupid Leo' on account of his successful sallies against the Bulgarians,[292] while Arab sources report that the men of al-Hadath only made a sally to capture some Byzantine siege engines after Say al-Dawla had marched to their relief (955).[293] What is more, the fact that cities and fortress like Castrogiovanni (835) were sometimes captured after failed sallies underlines the risky nature of this tactic.[294]

Although siege engines posed a significant threat to the security of a city, the most efficient way to breach the fortifications was to dig and undermine the foundations of the walls. This method was not only very effective, it was also quite cheap, requiring few resources and little specialized knowledge. To tackle this serious threat, the manuals recommend the excavation of a deep trench in front of the city 'because in this way, the enemy who is digging the tunnels will be clearly spotted'.[295] According to the *ST*, an alternative way to counter enemy mines was to fix metal objects into the ground so that the defenders could place their ears on the metal objects and hear whether somebody was digging underground. After the location of the enemy mine had been confirmed, the defenders were instructed to dig a counter-mine which would allow them to meet the attackers and neutralize their threat by suffocating them with gusts of smoke produced by burning felt and feathers. The smoke was to be directed to the enemy through a pair of bellows, similar to the type used by blacksmiths.[296] Nevertheless, the *ST* is not the first source to record this advice. Similar information survives in other military manuals as well: the *PS* advises the defender to place his ear into the ground so as to investigate if digging is taking place underneath, but also to counter-dig and confront the enemies underground by using smoke or by flooding the tunnel with water. In addition, parallel information is found in Heron of Byzantium and in the *De Obsidione*.[297] The oldest reference, however, to such course of action derives from Polybius' description of the siege of Ambracia by the Romans (189 BC).[298]

Given that this is a practice inspired by a classical source, it is reasonable to wonder whether it was applied on the Byzantine battlefield or whether it served as a theoretical and educational exemplar of the past. Byzantine historical narratives are silent about this specific tactic, except for the *Alexiad* which gives a detailed account of something similar. Anna Komnene informs us that during the siege of Dyrrachium (1081), the Normans dug a tunnel in order to undermine the walls. The Byzantines responded by digging a counter tunnel in which the men posted themselves to find out the exact location that the enemy was aiming at. When they managed to locate their direction through the sounds of digging, they opened small peepholes into the ground, and when they acquired visual contact with the enemy, they placed reed pipes through the holes and cast fire in the enemies' faces.[299]

Despite the fact that one may blame Anna Komnene as trying to portray the Byzantines in an ideal light here, her testimony could have been based on eyewitness accounts. Even though we must not forget that such accounts are also subject to distortion, the idealization of the Byzantines would have been achieved more effectively by drafting a narrative more identical to the advice of the manuals.[300] Alternative Byzantine and Arab sources seem to imply that the recommendations of the manuals were not totally alien. For example, Theophanes Confessor reports that in one of their raids (775–776), the Arabs used smoke to compel the fugitives of Cappadocia to come out of a cave, and al-Tabarī refers to the deep trench excavated by the Byzantines in front of Amorion, as well as to the difficulties that it caused the besiegers.[301]

To conclude, the advice of the *ST* on defensive siege warfare appears to be practical and relevant to a large extent.[302] Most measures find parallels in Byzantine, Arab and Western histories as well as administrative documents. The recommendations of the *ST* which find no mention whatsoever in other sources or are only recorded in historical narratives which aim to promote the deeds the military aristocracy remain problematic, like the use of pikes against battering rams.

Practical information 6: offensive siege warfare

The recommendations of the *ST* concerning offensive siege warfare can be divided into two categories: advice on how to win a siege with minimum fighting and taking a city by storm. The bloodless approach involved persuasion and famine. For both methods the author of the *ST* based his instructions on Onasander. To persuade the besieged, the general was to send ambassadors to the city, explaining that he is intending to ravage the countryside and inflict loss of food and income. The general was then encouraged to proceed with the implementation of his threats so as to compel the city to come to terms out of fear. An alternative route to persuasion was through mercy. The *ST* explains that if the general is humane to the cities which surrender to him, others may follow their example, expecting a similar treatment.[303] When mercy and threats did not work, the general often played the card of famine.[304] To compel the cities to surrender due to lack of food, the besieger was to 'assign himself the task of always hindering any interaction with the outside' and to 'capture those who go out for the collection of necessary provisions by mounting ambushes'.[305]

While these measures seem practical and standard, it is rational to wonder whether they reflect an outdated practice and whether they were indeed followed by tenth-century Byzantine generals. This task can prove tricky though, because to evaluate our sources correctly, we have to properly understand their weaknesses first. Generally speaking, most of the sources do not speak of negotiations or their details; they only report that the generals ravaged the countryside of a city without giving more information about their motives, being that famine, persuasion or booty.[306] When more detailed narratives are available, different problems arise. Take, for instance, the generalship of Nikephoros II Phokas. Leo the Deacon reports that Nikephoros II applied the measure of blockade at almost all his sieges. Thus, we learn that in Chandax he surrounded the city and tried to cut off its food supply, in Tarsus 'he encircled the town with diligent guards' trying to 'deliver the city into the grip [of] famine' and in Antioch he ordered his men to camp outside and to force the city to surrender by stealing its provisions with daily raids.[307] It is difficult to take this information at face value when almost every time the same sequence of events follows. To make matters worse, Leo the Deacon does not only portray Nikephoros II as following the instructions of the manuals, but he also presents him as giving similar advice in a fictional military oration full of rhetorical elements which was supposedly delivered in front of Antioch.[308]

John Skylitzes agrees with Leo the Deacon that Nikephoros II blockaded Antioch and disrupted food supplies.[309] His testimony, however, is questionable.

Even if we overlook the fact that he is a more remote source from these events, it is characteristic of him to standardize his battle descriptions and to report that most of the time cities surrendered because of lack of supplies.[310] This leaves the historian to wonder whether he comes across a clichéd phrase or an actual description of events. A good example is the capture of Dorostolon. John Skylitzes mentions that John Tzimiskes tried to capture the city 'by blockade and famine', whereas Leo the Deacon reports that the city was taken with a first assault.[311]

Despite the difficulties of our most detailed sources, readily declaring the information of manuals as null and impractical would be quite farfetched. More neutral sources like Byzantine chronicles and Arab historical narratives seem to imply that the blockade of a city was a standard procedure. We read, for instance, that Symeon blockaded Adrianople in 923 and managed to force a surrender because the defenders ran short of food and that Melitene surrendered to John Kourkouas because it was pressed by hunger.[312] Similarly, both Byzantine and Arab sources imply that the policy of threat and humane treatment was central to Byzantium's strategy. Historical narratives record that a relatively small number of cities or fortresses were captured by storm – most of them decided to subject themselves to the Byzantines by coming to terms with them.[313] The testimony of Arab sources is indispensable here as it supplements our knowledge of how cities were persuaded to negotiate and the way they were treated. We learn that John Kourkouas first violently ravaged the countryside of Melitene (934), whereupon the citizens decided to come to terms with him and finally opened their gates. The population was treated relatively leniently, as John declared that any Muslim who was willing to convert to Christianity would recover his family and his property, whereas those who chose to remain Muslims would be driven out of the city and moved to another region, but they, too, would remain unharmed and free.[314] A similar treatment is also recorded for the case of Tarsus which was captured by Nikephoros Phokas in 965.[315] The fact that this approach was mainstream and even encouraged by the capital is confirmed by legal sources. Romanos I used one of his novels as a stage to propagate this strategy. Romanos I took pride in that 'towns and cities have, with the help of God, come into our hands from the enemy, some as a result of war, while others have passed over to us by the example [of conquered towns] or through fear of capture'.[316]

In case the strategy of persuasion and famine was either not applicable or not severe enough to make a city surrender, the Byzantines employed a variety of siege engines to achieve their goals. The author of the *ST* draws on earlier manuals and devotes a whole chapter on how to storm a city or a fortress with siege engines and on which tactics to employ.[317] With regard to tactics, we read that the army should be 'divided into *allagia* (. . .) so that it does not easily become exhausted from the siege by fighting all at once. Having the greatest part of the army with him, the general ought to launch assaults continuously in relays, by night as well as by day', for 'continuous assaults will make the enemy weary and more likely to surrender'.[318] The most dangerous assaults were those undertaken during the night. The *ST* comments that the most appropriate time to attack is 'especially during the night, when the assaults seem a great deal more intimidating to the besieged'.[319]

Evidence from Byzantine, Western and Arab sources seems to agree that the advice of the *ST* was still relevant and practical. Liudprand of Cremona highlights how dangerous night attacks were and records in detail the measures taken to guard Constantinople against similar hostile actions at the time of Leo VI. According to Liudprand, patrols were stationed in short intervals all around the city with the duty to arrest, to interrogate and to deliver for public trial anyone who they encountered during their watch.[320] An eyewitness to the siege of Thessaloniki, John Kaminiates shares that a potential night attack caused significant fear and anxiety to the defenders of the city. The Arabs, we learn,

> fought there until late into the night and then (. . .) rested (. . .). Though perhaps they were exercising their minds how best to attack us (. . .) and were intent on preparing a further series of treacherous and deceitful moves. No sooner, therefore had we paused a moment from the heat of battle than we were thrown into a further state of anxiety over the level of vigilance maintained by the troops manning the fortifications that ringed the city and the suspicious movements of the barbarians, movements which might be the prelude to a successful ambush carried out under cover of darkness that would allow them to penetrate our defences undetected and thus encompass our destruction (. . .). Accordingly, we stayed awake all that night.[321]

Last but not least, al-Tabarī records that the Arabs tried to take advantage of the physical and psychological fatigue of the garrison of Amorion. The Arabs attacked in relays, day and night, believing that this would force the Byzantines to surrender. In the first stage of the siege, however, the Byzantines responded well and managed to withstand by guarding the towers in rotation.[322]

The success of the assault depended, however, not merely on time and persistence but also on the employment of siege engines. The *ST* encourages the use of πετροβόλων ὀργάνων and ἐλεπόλεων which denotes stone-throwing machines.[323] Reference is also made to portable wooden siege towers, to battering-rams, to tortoises and to wheeled ladders.[324] The use of some of these siege engines is well documented. The most devastating, the stone-throwing machines, were regularly employed in the tenth century.[325] The term ἐλέπολις had a more generic meaning in the middle Byzantine period; it could denote anything from a siege tower, as it was its original meaning in antiquity, to a ram and a stone thrower.[326] It is clear, however, that in the context of the *ST* reference is made to a certain type of stone thrower, which must have resembled the trebuchet. Perhaps the use of both terms denotes two different types of machines with different capabilities in terms of range and velocity.[327] The wheeled ladder appears to have been a Byzantine innovation since its existence is not recorded in any source from antiquity.[328] Byzantine histories refer to the use of ladders during sieges without specifying their type though.[329] Nevertheless, we have a clear mention of wheeled ladders in the seventh century, as they are reported during the siege of Thessaloniki by the Avars and Slavs, as well as in the tenth century, in the poem of Theodosius the Deacon which mentions 'composite ladders', a type which appears to have been similar to the wheeled ladders found in the manuals.[330]

On the other hand, other siege engines are only widely mentioned in military treatises. Devices like the wooden siege towers, the battering rams and the tortoises are mentioned both in specialized and more generic military treatises. Generic military manuals like *ST*, *LT* and *TNO* treat these machines without many details, which implies that they were already well known and probably continuously used throughout Byzantine history.[331] In turn, treatises specializing in siege warfare, like that of Heron of Byzantium, describe in detail many sub-types of these machines, like a special tortoise for the undermining of the walls called χελώνη ὀρυκτρίς. Given that these machines were already employed from antiquity and that specialized manuals describe devices that are never found in histories, it is rational to wonder whether these are antiquarian references, whether the Byzantines actually used such devices and whether these machines were evolved through time.[332]

A characteristic example is the case of the battering ram and the tortoise. The most contemporary and detailed account we possess for the use of these devices is the testimony of Leo the Deacon. In his history we read that Nikephoros II Phokas employed a ram during the siege of Chandax (961). Leo the Deacon comments that 'this is the device the Romans call a battering ram, because the piece of iron that is joined to the beam and batters the town walls is shaped like a ram's head'.[333] At the same siege we are informed that a tortoise was used to undermine the walls of the city. Leo the Deacon describes how the Byzantines 'carrying stone cutting tools began to dig there quietly chipping away and cutting through the rock at the point where the foundation of the wall was set', while a very similar course of action is given for the fall of Mopsuestia.[334] The appearance of almost identical testimonies for two different sieges obliges the historian to approach this information with caution. Leo the Deacon may have not recorded real events but could have somehow elaborated his account. This becomes more of an issue since we know that the history of Agathias was a model for Leo the Deacon and that these accounts resemble the siege of Cumae (552). Leo the Deacon's dependence on Agathias also leaves us wondering whether this is indeed a contemporary description of a ram or an act of imitation.[335]

Even though Byzantine historical narratives are sometimes problematic, there are a number of alternative and reliable sources which demonstrate that these siege engines were not a purely literary creation, but rather a standard piece of equipment in the arsenal of Byzantine generals. Ibn al-Athīr reports that John Kourkouas used siege towers during the siege of Dvin (927), and the administrative lists in the *DC* include a wooden tower among the equipment prepared for the Cretan expedition of 949.[336] These lists, uninfluenced by literary conventions, provide a safe testimony and record, among others, ten rams, as well as tortoises, together with sledge hammers and pickaxes, most probably for undermining purposes.[337]

Siege engines required expertise, resources and time to be produced. It was therefore essential to protect them from enemy missiles and attacks so that all those resources and time should not be wasted. The *ST* responds to this problem by advising the general that siege engines

> should be enclosed all around with newly stripped-off buffalo hides. The tortoises should be smeared all over with clay on top and sponges completely soaked with vinegar should be placed on their exterior. For if vinegar is used, it prevents the combustion of fire, especially of the so-called liquid [fire].[338]

Specialized manuals on siege warfare like that of Heron of Byzantium and more generic ones like *LT* include similar instructions.[339]

These measures were probably quite mainstream, as Byzantine and Western historical narratives record that they were widely employed in the east. Michael Attaleiates, for instance, states that the Turks covered their stone-throwing machine with fabrics which could deflect missiles but proved to be vulnerable to fire.[340] John Kaminiates most probably refers to some similar protection when he records that the Arabs 'ἑπτά παρέστησαν πετροβόλους πάντοθεν περιπεφραγμένους'.[341] Albert of Aachen reports that the Crusaders were advised by local Christians to use vinegar in order to counter the attempts of the garrison of Jerusalem to get the better of them through the use of fire. The administrative lists of the *DC* corroborate our knowledge, recording that sponges were included as parts of siege engines in the campaign against Crete (949).[342] Negligence in these matters could prove disastrous for siege operations. Most probably the Byzantines failed to cover their siege engines during the siege of Serdica (986), and Leo the Deacon reports that 'after the siege machines and the other contrivances accomplished nothing, because of the inexperience of the men who brought them up against the walls, they were set on fire by the enemy'.[343]

Although siege engines were indispensable to destroy or to overcome the walls, the siege was by no means over after part of the walls had been demolished or captured. It is safe to assume that once the walls had fallen, the garrison would resist more daringly out of desperation for the safety of their lives and of their families. The same point of view seems to be confirmed by Byzantine and Arab historical narratives which reveal or imply the dangers that the besieger could face when he entered a city. We know, for instance, that the Byzantines managed to get close to the gates of Melitene, only to be driven back by stubborn resistance, and that they were repulsed at the siege of Dvin right after they had breached the walls.[344] Eyewitness accounts, like that of John Kaminiates, record the same sentiment. In John Kaminiates' words, once the Saracens approached the walls, the garrison of Thessaloniki 'came to think nothing of death, since it was both inevitable and staring them (. . .) in the face'. Thus, they 'threw themselves unreservedly into the struggle, making the moment of maximum danger an occasion for displaying their courage (. . .) every man did his utmost'.[345] John Kaminiates also explains that fighting through the narrow streets of a city could prove a challenging and dangerous task for the attackers. More specifically, he reports that after the Arabs had captured the battlements of the city, they:

> waited for the crowd to surge forward, trying to discover whether they had made off in feigned or in genuine flight. For they suspected that the inhabitants might have laid some hidden ambush for them in the streets, in order to way-lay them once they had split up into separate groups. Consequently, they were reluctant to enter the city, and set about their task without first taking precautions.[346]

The *ST* makes explicit reference on how to tackle these last-stand situations. The general is instructed to

> announce in the language of the enemy (. . .) that none of the citizens who are unarmed are to be slain (. . .). For if the enemy hears such an announcement, everyone (. . .) will shamefully choose slavery, and so, when the city becomes empty of armed men, it will be captured without danger.[347]

Thus, the author of *ST* proposed the avoidance of further fighting after the walls have been breached, and this advice is given purely on utilitarian grounds so that the city will be captured without risking extra casualties.

Promotional sources, however, give a different dimension to this instruction. Leo the Deacon records the way in which Nikephoros II Phokas acted after the walls of Chandax had been finally breached (961). We read that the Arabs:

> turned to flight, withdrawing through the narrow streets, as the Romans pursued and slaughtered them mercilessly. The survivors, and those whom the warfare had not succeeded in mowing down, threw down their arms and turned to supplication. When the general observed this, he (. . .) restrained the soldiers' onslaught, persuading them not to kill the men who had thrown down their arms, nor to attack cruelly and inhumanely men without armour or weapons, saying it was a sign of inhumanity to cut down and slay like an enemy men who had given themselves up in surrender.[348]

It could be that Leo the Deacon wanted to idealize Nikephoros II Phokas here: the latter follows the advice of the manuals not to harm unarmed people and to avoid further fighting, but the purpose for this course of action was neither utilitarian nor necessary, merely the choice of a pious man. Thus, Nikephoros II Phokas emerges as both a successful and merciful general.[349]

In conclusion, the *ST* seems to provide us with more or less practical advice regarding offensive siege warfare. In spite of the fact that some of our most detailed sources are problematic and include accounts which seem to be generic, manipulated, elaborated or even a product of imitation, the testimony of alternative Western, Arab and Byzantine historical narratives in connection with administrative and legal documents corroborates our knowledge and seems to agree that the manuals touched upon practical issues.

Practical information 7: numbers, units and battle formations

Any attempt to evaluate the numbers of troops given in the *ST* comes up against the problem of how to estimate the numbers of the Byzantine army accurately in the first place. This issue is controversial and has already been discussed in great detail and at great length by other scholars. There is therefore no need to repeat their findings here.[350] In order to estimate how realistic the numbers in the *ST* are, there is no need specifically to debate the total numbers of the Byzantine army, but rather to look at how large an expeditionary army could be, comparing the information of the *ST* with other sources. Since historical narratives offer controversial and sometimes exaggerated numbers, it would be better to take into consideration other types of accounts as well, such as the *DC*, which contains administrative lists, and the testimony of Arab sources.

The *ST* provides the reader with a number of variations of battle formations, each for a different number of men, consisting exclusively either of infantry or cavalry or of both. The highest number that the *ST* gives is a combined army of 26,184 men, 19,414 of whom were infantry and 6,770 cavalry.[351] To find out whether the Byzantines could mobilize such a force in the tenth century, one should first turn to the administrative lists of the *DC* which seem to have fulfilled a utilitarian purpose and appear to be untouched by biases or exaggerations. The *DC* seems to agree that the highest figures of the *ST* were a realistic representation of the tenth-century Byzantine manpower. It informs us that, theoretically, in the Cretan expedition of 911, approximately 37,000 men were mobilized, around 6,000 of whom were cavalry, while for the expedition of 949, around 12,600 men are recorded, around 6,000 of whom were cavalry.[352] What is more, this evidence is backed by Arab sources which provide similar estimates of Byzantine forces, a good example being al-Athīr, who reports that the expeditionary force of John Kourkouas against Melitene in 934 comprised 30,000 men, and Leo the Deacon, who states that 28,000 men campaigned with John Tzimiskes.[353]

If the highest figures of the *ST* were practical indeed and Byzantium could afford to muster an infantry and cavalry army of 26,184 men, there seems no reason to doubt that the smaller variations of 12,528, 9,220 and 4,116 men could have reflected contemporary practice. As a matter of fact, these numbers do not seem to have been proposed at random, since their frequent occurrence in manuals of war suggests that they were more or less mainstream. For instance, the *DRM* considered 12,000 men an adequate operation force. *LT* speaks of an infantry and cavalry army of 12,000, 10,000, 5,000 and 4,000 men, while a ninth-century Arab manual states that 12,000–4,000 men was a serviceable number for military operations.[354] Similar observations can be made for the purely cavalry formation of 6,770 men the *ST* speaks of.[355] This number is in line with the advice of other manuals as well, like the *PM* and the *DV*, where we read that 'who has five or six thousand warlike horsemen and the assistance of God will not need anything more' and of a cavalry force of about 6,000 men.[356] The account of the *DC* records twice a cavalry force of around 6,000 men. On similar grounds, the other two smaller variations found in the *ST*, of 3,000 and 1,000 cavalrymen, respectively, also seem to have had an operational value. The *DV* states that a *thema* could have had 3,000 cavalrymen, and the observations of scholars about large *themata*, like that of Thrakesion in the tenth century, match this number. Smaller *themata*, like that of Peloponnesus, appear to have had the means of providing the expenses for 1,000 cavalrymen, and it seems that such a number was also operational, as the *DC* records that Romanos I sent a cavalry detachment of 1,453 men to Italy in 935.[357]

While these numbers and variations were most likely realistic and practical, other figures appear to be quite ambiguous. For example, the *ST* records a purely cavalry army of 18,570 men. Compared to other manuals, this figure is quite large – probably the closest example we have is the testimony of *LT* which speaks of a cavalry army of 12,000 but nevertheless comments that the basic one consisted of only 4,000 men.[358] Unfortunately, it seems that we do not have secure and precise information from alternative sources which argues that such a number of cavalrymen was regularly deployed. Perhaps the 18,570 reflected a figure which was in accordance with theoretical registries of Byzantine *stratiotai*, which seem to have been outdated and not to have reflected the actual number of able cavalrymen who could have participated in an expedition.[359]

Equally problematic is the purely infantry army of 24,100 men, but the ambiguity in this case is not so much in the numbers.[360] In our discussion earlier, we have seen that the *DC* records compound expeditionary armies of 37,000 and 12,600 men; if we subtract the cavalry from these numbers, we get two infantry armies of 31,000 and 6,600 men. Assuming that the cost of maintaining a marine or a ship crew would have been more or less the same as that of a medium and of a light infantry man (in the *DC* the men are recruited from nautical *themata*), it seems that the number of infantrymen the *ST* reports is realistic. The *ST*, however, records that these 24,100 men were exclusively infantry and supposed to act on their own accord independently of cavalry. It seems that we do not have any reference from historical narratives to determine whether this was actually the case, and the lack of sources on the Byzantine infantry in general makes the issue all the more uncertain. Be that as it may, one can only compare the advice of the *ST* with other manuals. The closest narrative to that of the *ST* appears in *LT*. Leo VI describes a force of 24,000 infantrymen, who are in similar array to the one given in the *ST* but who nevertheless operated with cavalry and not on their own. The testimony of later manuals preserves nothing similar to the advice of the *ST*. For instance, the *PM* speaks of an infantry army of 12,000 men who, once again, operate with the help of cavalry and are drawn up in a hollow-square formation.[361] Whatever the case, the lack of parallel evidence can be explained by the fact that the array of the *ST* may not have been intended so much for regular combat, since it could also be adopted by a compound army in which the cavalry dismounted in order to cross defiles of medium width.[362]

But numbers aside, we also possess very little evidence to determine whether these infantry or cavalry formations were ever employed by generals on the battle-field. Historical narratives are usually of little to no help since they rarely comment on how the army was arrayed. What we usually get from them is the almost standard phrase that the commander divided his troops into three divisions: left, right and centre. Even when we get a little more than that, the description is given in passing, leaving the historian uncertain whether this information reflects a tactic recorded in the manuals or not. For example, the *ST* states that the standard battle array for the infantry was the hollow-square formation, which, according to circumstances, could also be either a horizontal or a vertical rectangle.[363] In turn, Leo the Deacon records that Nikephoros II Phokas arranged his army in an oblong formation during the siege of Chandax but gives no further detail on whether this was a simple classical-style formation or whether it was a hollow one.[364] Surprisingly enough, military manuals are no better, since when they refer to the Arabs, they only record that their 'battle formations are both square and oblong and so are very secure' and that 'they also imitate the Romans in many respects', without giving any more detail.[365]

The exceptions to the rule are a few narratives which describe infantry and cavalry formations more extensively. For example, Anna Komnene speaks of a formation that was supposedly invented by Alexios himself. The formation was a hollow square in which the booty, rescued prisoners, women and children were placed on the inside. Anna describes Alexios' hollow square as 'a moving city' which protected everybody and was so perfect and solid as if 'it was directly attributable to God' and to the angles. She comments that 'the ranks' were 'organised in such a way that the Turks would have to shoot from their right at the side protected by the shield, whereas our soldiers would shoot from the left, that is at the side that is unprotected'. Anna continues to report that 'the serried ranks of close-locked and marching men gave the impression of immovable mountains' and that the men marched and acted as one body 'to the sound of the flute', marching slowly 'at an ant's pace'.[366] The same author, along with Nikephoros Bryennios, records that Alexios I and the homonymous grandfather of Bryennios fought in the battle of Kalobrye (1078), dividing their armies into three main units and posting flank guards and out-flankers in accordance with the suggestions of manuals. The battle of Kalobrye preserves another parallel deployment with the manuals since Alexios I is stated to have used concealed troops which he placed at the outermost left flank of his formation.[367] Leo the Deacon preserves a detailed description of the wedge formation of the *kataphraktoi*. As per the advice of manuals, Nikephoros II Phokas deployed his heavy cavalry supported by horse archers outside of Tarsus. We read that he attacked, 'deploying the ironclad horsemen in the van, and ordering the archers and slingers to shoot at the enemy from behind'. Leo the Deacon also comments on the gleam of the heavy cavalry's armour and on the discipline of the assault, and finally reports that the Tarsians were 'forced back by the thrusts of spears and by the missiles of the [archers] shooting from behind'. Finally, a similar course of action, we learn, was undertaken by John I Tzimiskes, who 'deployed the Romans in the van and placed ironclad horsemen on both wings, and assigned the archers and slingers to the rear and ordered them to keep up steady fire'.[368] According to John Skylitzes, John Tzimiskes seems also to have followed the recommendations of the manuals regarding the use of scouts. We learn that he 'detached a company of picked men (. . .) with orders to advance ahead of the army, look out for the main body of the enemy and to keep the emperor informed. If they drew near they were to test the strength of the enemy by skirmishing them'.[369] This course of action brings the advice of the *ST* to mind, which recommends the use of a unit of scouts (*prokoursatores*) to reconnoitre. These men were to attack the enemy with arrows, and if they were unable to overcome him, they retreated behind the vanguard, which undertook the attack.[370]

Although these are our most detailed accounts, their information cannot be taken at face value as they are biased and problematic. Anna Komnene appears to be keen on enhancing the image and prestige of Alexios I. She states that the hollow-square formation was a novelty, but we do know that such formations existed long before her father. On this matter she seems to contradict herself to a certain extent because she says that Alexios drew this formation on paper because he was familiar with Aelian's tactics. Given that there is no other account to confirm whether Alexios employed such tactics, it would be dangerous to take the testimony of the *Alexiad* at face value. Anna presents it in such an ideal light that it is impossible to accept her narrative. She records that the square halted every time a woman was giving birth, or someone needed medical help, or somebody died and a burial was required and that the emperor was present at all these and personally cared for them. It could be that Anna Komnene knew of the interpolated version of Aelian, which also included the *Syntaxis Armatorum Quadrata* that preserves the diagram of a hollow-square formation. This diagram could have been the inspiration for Anna Komnene to model her account accordingly and to manipulate events with a view to presenting Alexios as somebody who could protect his people, who could easily win the battle and provide care as well as refuge to every single soul.[371] Anna's narrative of Kalobrye is also suspect, although she did not copy Bryennios' history word for word; both historians had good reasons to portray their subjects as playing it by the book. Since Anna Komnene promoted the cunning side of Alexios I, and Bryennios the courage and good order of his relative, the exact same account could be used to convey a different moral message. The use of concealed troops by Alexios I especially could have served to contribute to the image of the master trickster Anna Komnene was trying to build for her father. Not only were the concealed troops deployed at the outermost flank of the formation, where the *ST* puts them, they are also recorded to have had a crucial effect on the enemy's morale, much like the treatises of war suggested. We read, therefore, that Alexios' concealed troops almost turned the whole of Bryennios' right wing to flight.[372]

As far as Leo the Deacon is concerned, his account seems to be so close to the advice of the *PM*, one could take it at face value and argue that Nikephoros II Phokas and John Tzimiskes used military manuals to fight their wars.[373] However, taking into consideration that Leo the Deacon is very favourable to these two generals, it seems very possible that he tried to idealize him. In fact, what better way to eulogize Nikephoros II Phokas than drafting a battle narrative in which Nikephoros II defeats the enemy with the same advice that he gives in his manual, the *PM*?

John Skylitzes' reference to the use of scouts by John Tzimiskes is problematic on the grounds that we know his narrative is infused with promotional lost sources. It is perhaps no coincidence that these details go hand in hand only with famous and eulogized figures, a good example being that the *Vita Basilii* recounts that Basil I took the same precaution since he 'set aside a body of picked men whom he sent forward as scouts and reconnoitres; he himself then followed with his main force'.[374]

Although we cannot too readily accept the information of these narratives, caution is somehow mitigated when a different reading is applied to these sources and their evidence is put together with Arab and Western material. For example, Anna Komnene undoubtedly exaggerated and glorified Alexios' formation, but the latter could have indeed been employed in practice. It is certainly not impossible that an experienced general like Alexios knew of and used such a formation. Alexios I might have had access to military writings and perhaps to the interpolated version of Aelian. He could have been inspired by the diagram of the *Syntaxis Armatorum Quadrata*, and he could have deployed his troops in a similar manner. In fact, Western sources speak of a square fighting-march formation which was employed by the Crusaders at the battle of Ascalon (1099). This was a square which was formed by nine units in total, three in the front, three in the back and three in the middle. Although this square was simplified and not hollow, it employed all men by ranks, just as the one in the *ST*. As this tactic was unprecedented in medieval warfare, one may entertain the possibility that it was inspired by Byzantine tactics.[375] If this is indeed the case, then the square of the *ST* seems to be the closest candidate to have inspired the Crusaders' formation. Similarly, our evidence for the battle of Kalobrye could have reflected the actual strategy of Alexios I and Nikephoros Bryennios. This becomes more evident once we notice that Anna Komnene's and Nikephoros Bryennios' narratives do not follow the advice of manuals word for word. Instead of playing it by the book and posting both flank guards and out-flankers, having numerical superiority, Bryennios only posted out-flankers, and Alexios I only flank guards, while the latter, instead of concealing one unit at each flank of his formation, only concealed troops at his left flank. The fact that both commanders did not adhere to the advice of manuals verbatim makes the accounts more realistic and implies that the reality of the field dictated that sometimes the recommendation of the manuals be followed only partially depending on the situation. What is more, Arab sources confirm that such measures must have been more or less standard, or perhaps introduced by the Byzantines as a counter-measure. From the military dispatch of Muḥammad Sulaymān (903) we learn that the Qarmatians used such tactics since they concealed two units, one in their left and one in their right wing, which they used to ambush the enemy.[376]

On closer look, the history of Leo the Deacon becomes a little more plausible too. Leo the Deacon seems to have completed his history after the *PM* was written, but instead of drafting the charge of the *kataphraktoi* on this treatise, his battle description follows more closely the advice of the *ST*.[377] In the *PM*, the *kataphraktoi* use the mace as their primary weapon, and the wedge of the *kataphraktoi* is always posted in the vanguard. It is only in the *ST* that we find the *kataphraktoi* using the spear as their primary weapon and the wedge being drawn up either in the centre of the vanguard or in the two flanks, as was the case with John Tzimiskes.[378] The fact that Leo the Deacon's narrative does not follow verbatim the manual of the figure it is attempting to eulogize could denote that his description was based on actual Byzantine tactics. After all, some accept that Leo the Deacon had access to oral sources. His source could have been an eyewitness and a high-ranking officer who saw the advice of the *ST* employed on the battlefield and then reported the events to Leo the Deacon, undoubtedly in a sympathetic light.[379] Whenever we possess alternative accounts, the advice of the *ST* concerning the *kataphraktoi* and the narrative of Leo the Deacon seem to be confirmed. Arab sources demonstrate that the equipment of the *kataphraktoi* was far from a mere literary imitation designed to bring to mind the heavy cavalry of antiquity. An eyewitness to the Arab–Byzantine conflicts of our period, al-Mutanabbī, provides us with an account of the Byzantine *kataphraktoi*, recording that they were completely enclosed in iron, bearing iron garments and helmets, while their horses seemed to have no legs, most probably due to horse armour.[380] What is more, the armament of the *kataphraktoi* in the *ST* (heavy armour with bows, lances and shields) seems to have been the standard way to equip heavy cavalry and probably reflected either mutual influence or a response to enemy developments, since according to al-Mas'udi (896–956) the Khazars were 'heavy armed mounted archers with bows, lances and shields like all the rest of the Arabs'.[381]

Finally, although John Skylitzes' account is very similar to the advice of the manuals and this tactic is mostly recorded for favoured figures in Byzantine histories, it could be that this was supposedly a standard tactic. In fact, Leo the Deacon also records that John Tzimiskes used scouts in this case, and the two narratives do not seem to follow one another closely.[382]

In conclusion, it is quite problematic to determine how practical the battle formations of the *ST* are. As far as numbers are concerned, most of them seem to be quite realistic, as they are confirmed both by the *DC* and by moderate estimates in the historical narratives. Some of the numbers, however, like that of the purely infantry formations, remain problematic, given that the information we possess is very scarce. As far as the actual formations are concerned, we are once again limited by our sources. Most of them are of no help, while others, although they provide more detail, cannot be accepted at face value, because they are promotional. In cases where only such narratives exist, one should be cautious not to accept or discard their testimony too readily – that is before appreciating their complexity. It seems safe, however, to accept that certain aspects of the advice of manuals on formations were indeed employed in action, especially when they also appear in Western and Arab accounts.

Practical information 8: guerrilla warfare

Despite the fact that the Byzantines had already started to take the offensive against the Arabs before the middle of the tenth century, guerrilla warfare remained important almost throughout the century and was very effectively employed by small armies of the *themata* in order to defend their territory against Muslim raids or expeditions. We have already seen that the *ST* covers such matters with some detail. The relevant information of the *ST* can be grouped into two general categories: first, advice on how to protect against enemy guerrilla tactics when an invasion or raid was undertaken, and second, advice on how to employ guerrilla tactics to counter enemy raids or invasions.

With regard to the first category, the *ST* instructs the general to be cautious when raiding in hostile territory and to pay close attention to narrow passes because 'the enemy has only to pre-empt the defiles and to rekindle hostilities during our withdrawal, when our army [is] burdened by the spoils. In addition, it [is] likely that the enemy will emerge victorious since they fight fresh and on ground to their advantage against those who are weary'.[383] Given that the enemy was in an advantageous position once he had captured the defiles, the manuals recommend not to 'join battle in the defiles' and not to 'force the way through, because this is, in fact, the most dangerous course of action'. Instead, they advise to retreat from the defiles and by ravaging the nearby regions, to lure the enemy out in more favourable terrain.[384] The best way to avoid such misfortunes, however, was to ensure that the enemy did not capture the defiles. The *ST* proposes to occupy with infantry the defiles from which the army will enter and exit enemy territory, so as to prevent the enemy from blocking and ambushing the Byzantine force.[385]

Even though the advice of the treatises seems to have touched upon very practical issues, it is best to look at our sources critically and closely before determining whether their advice was actually employed in the battlefield. For example, the practice of avoiding giving battle in defiles once trapped into enemy territory is hardly ever recorded in Byzantine historical narratives. One of the few exceptions is the narrative of John Skylitzes. According to the *Synopsis Historion*, in 1014 the Bulgarians blocked the defiles and trapped Basil II and his forces. Basil II did what the manuals advise not to do and attempted to force his way through the blockade, which resulted into a large number of casualties for the Byzantines. Nikephoros Xiphias, one of Basil II's generals, offered to find another way through the defiles and then surprise the enemy by attacking simultaneously with the emperor, which was how the battle was won.[386] Given that John Skylitzes based some of his passages on sources which were sympathetic to the military aristocracy, it is hazy whether this account was set to portray Nikephoros Xiphias as a capable general who confronted the situation, acting in accordance with the manuals.[387]

Similarly, we learn from the *Vita Basilii* that after the Byzantines had attacked Adata, they posted various detachments to capture the defiles because they expected some kind of attack during their retreat. While this narrative is very close to that of the manuals, the purpose of the *Vita Basilii* was promotional, which makes this passage difficult to accept at face value. Arab sources suggest that Sayf al-Dawla followed a comparable course of action. We read that in 951 he guarded all the passages behind him in order to ensure the safe retreat of his army from the raid. Although this alternative account could have been of help here, its evidence is not conclusive, since it appears after Sayf al-Dawla's significant defeat from the Byzantines in 950 and could have served as an offset to show that he learned from his misfortunes.[388]

Nevertheless, even if these accounts were deliberately angled to idealize their figures, our knowledge is supplemented by more neutral accounts which imply that these measures were commonplace. To begin with, the testimony of Byzantine, Armenian and Arab sources confirms that the narrow mountain passes were often captured so as to trap and easily defeat the Byzantines. Leo the Deacon, who was an eyewitness to this event, reports that in 986, after the siege of Serdica, the Byzantines were ambushed by the Bulgarians in a defile. The army was annihilated and the enemy captured the imperial baggage train, something which is also mentioned by Stephen of Taron.[389] Arab accounts record that after the Byzantines had captured and abandoned Samosata (927), they were attacked and defeated by the Arabs, who managed to take back a large number of spoils. We also read that Sayf al-Dawla managed to completely defeat the Byzantines in a defile and to reclaim the booty after Bardas Phokas was withdrawing from a successful raid against Marash (944).[390] Similar accounts imply that the rational measure of avoiding battle in the defiles was generally employed on the battlefield. We learn that when the Byzantines blocked the defiles to hinder Sayf al-Dawla's retreat, the latter responded by looking for alternative passages, making a wide detour and ravaging the enemy land on his way.[391] This narrative seems to be more credible due to the fact that Sayf's manoeuvres either had little effect or did not result in an immediate victory, as was the case with Xiphias. Last but not least, we also find a casual reference to the fact that Sayf al-Dawla attempted to capture the defiles with an infantry force only to find that they had already been occupied by the Byzantines (953).[392]

Whenever the Byzantines found themselves on the defensive, however, they employed similar guerrilla tactics to counter enemy raids and invasions. Following earlier Byzantine manuals, the *ST* advises that 'the general must not fight those who occupy our land openly and in a pitched battle but during the invasion he must not oppose them at all. However, he must always lurk in strategic locations'. The *ST* also comments that 'if the enemy is retreating and by this time has reached his own borders (. . .) then it may be advantageous to attack furiously, (. . .) because everybody is weary from the march and likewise encumbered'.[393]

There is no doubt that this strategy was widely employed by the Byzantines since these measures were by no means a novelty of the tenth century. Byzantine, Arab and Armenian sources agree that this was more or less the traditional Byzantine approach to warfare at least from the eighth century onwards.[394] What is more, with regard to the tenth century in particular, Arab sources provide a credible and casual account of more trivial events, which confirms the testimony of the manuals. For example, we know that when Sayf al-Dawla invaded and ravaged the *thema* of Koloneia (939), John Kourkouas only attacked him as he was retreating. Accordingly, in 956, John Tzimiskes protected his own *thema* as its *strategos* by avoiding a pitched battle with Sayf al-Dawla and instead captured the defiles to block his exit and ambushed him as he was withdrawing into Arab territory.[395]

By far the most perfect application of the instructions of the *ST* is found in how Leo Phokas dealt with the raids of Sayf al-Dawla in 950 and 960. These events are most interesting because they are recorded in three different sources: John Skylitzes, Leo the Deacon and Yahya of Antioch. Therefore, a comparable study of these narratives can shed significant light on the working method of Byzantine historians and demonstrate the extent to which they manipulated events to present certain figures in an ideal light. Our first testimony, John Skylitzes, reports how in 950 Leo Phokas was informed of Sayf's movements and ambushed him in a defile as he was retreating. Sayf al-Dawla 'was surrounded by forces lying in ambush. Men posted for this purpose rose up from their concealed positions, rolling great stones down on them and shooting all kinds of missiles at them'.[396]

Leo the Deacon goes a bit further than John Skylitzes and into far greater detail for the events of 960. As per the advice of the manuals which dictated avoiding battle in the absence of an able force, Leo the Deacon makes specific reference to the fact that Leo Phokas 'decided not to expose the army to certain danger, nor to face the barbarian host in the open'; instead, he proceeded to 'occupy the most strategic positions on the precipices, to lie there in ambush and guard the escape routes, and then confront the barbarians at the most dangerous and perilous sections of the path'.[397] Leo the Deacon's narrative is full of praise for Leo Phokas. In order to enhance Leo Phokas' image, Leo the Deacon drafted a fictional speech which was so perfect that when delivered to the army supposedly made all soldiers want to follow Leo Phokas to their death.[398] What is more, Leo the Deacon openly suggested that Leo Phokas was 'a courageous and vigorous man, of exceptionally good judgement, and the cleverest of anyone we know at devising the proper course of action at times of crisis', going as far as to say that 'some divine force, I believe, used to fight alongside him in battles'.[399] The sympathy of Leo the Deacon for Leo Phokas was not unprecedented – praise for the achievements of the same general is found in another pro-Phokas source, in the preface of the *DV*.[400]

Against this background, the historian would have wondered to what extent Leo the Deacon and perhaps John Skylitzes idealized the deeds of Leo Phokas by drawing upon promotional material or even by shaping their battle descriptions to match the testimony of the manuals.[401] But in this case, we are lucky enough to possess the alternative testimony of Arab accounts, which can determine to what extent the Byzantines altered their accounts. The sources report that in 950 Sayf al-Dawla returned from his raid in the *thema* of Charsianon only to find that the *kleisourai* were already captured by Leo Phokas and that the passage was blocked with tree trunks. The Byzantines attacked with missiles, and Leo Phokas attacked the rear guard of Sayf al-Dawla, who managed to fight his way through with considerable losses. Sayf tried to cross via an alternative route, marching through rough terrain.

His troops were already exhausted from the previous engagement, from the march and from carrying the spoils. Towards the end of their journey, close to the *kleisoura* that led to the southeast of Marash, the Byzantines attacked again and utterly destroyed his army.[402] With regard to the second victory in 960, we learn that on his retreat, Sayf al-Dawla was blocked in a defile by Leo Phokas who ambushed him and destroyed his army.[403]

The testimony of Arab sources agrees with the narrative of Byzantine histories in terms of the main strategy employed. Promotional Byzantine accounts therefore certainly praised and idealized figures, but in this case at least, they do not seem to have fabricated events out of thin air in order to glorify Leo Phokas. The fact that the engagements were fought near the borders of the enemy may have something to do with the advice of the *ST* which urges the general to attack the enemy in this location 'because they will be more negligent since they clearly have more opportunities to save themselves'[404] Indeed, Arab sources inform us that some of the enemy abandoned Sayf before the engagement, most probably to run towards their own territory and to ensure safety.[405]

Similar observations can be made about the way Byzantine historians recorded ambushes which were mounted deep into the Byzantine territory.[406] From the manuals we read that the general should conceal two units and use another one to lure the enemy towards the ambush with a feigned retreat.[407] Perhaps the most famous parallel account of this tactic is found in the history of John Skylitzes, since its narrative represents an almost perfect implementation of this advice, demonstrating how a general should deal with superior numbers and employ ambuscades.[408] We read that in 970, Bardas Skleros prepared an ambush against the Rus and sent John Alakasseus to lure the enemy with his detachment by feigning retreat.[409] Knowing that John Skylitzes has employed pro-Skleros material in his history, one cannot but wonder about the veracity of this passage.[410] Matters become less vague when we read the same events from the history of Leo the Deacon. The alternative account gives us a peephole view of the process of historical writing and the agenda of every author. Leo the Deacon and John Skylitzes only agree on the fact that the Byzantines mounted an ambuscade. According to who they wanted to promote, however, they disagree on who was responsible for coming up with the plan and who led the operation.[411]

The last tactic the manuals propose to deal with invaders on regular Byzantine territory is to surprise them when they were unprepared for battle and occupied with some other task. For example, the *ST* instructs the general that 'the very best time for attacking (. . .) is when the army is occupied with fixing the camp and whenever everybody releases the horses for pasturage'.[412] It seems that this tactic was indeed employed on the battlefield. All the chronicles agree that when the Rus invaded Byzantium in 941, Bardas Phokas attacked them 'when he encountered a considerable body [of Rus] set to forage'. His example was followed by John Kourkouas, who, when he arrived with reinforcements, 'found the Russians dispersed wandering hither and thither and dealt them a bitter blow'.[413] This strategy was not only employed by the Byzantines – it was more or less commonplace. We have already seen that Byzantine narratives record how such attacks were under-taken against the Byzantines as well, while Liudprand of Cremona records that the Hungarians employed the same practice in the west.[414]

To conclude, the plethora and variety of information we possess on guerrilla warfare seems to validate the practicality of the instructions provided by the *ST*. This is better demonstrated by the fact that guerrilla warfare was the basic strategy of Byzantium for almost three centuries and remained important during the tenth, despite the Byzantine offensive. Though with regard to tenth-century practice, we find parallel information in promotional narratives, their testimony seems to be supplemented and validated by more neutral Byzantine accounts, as well as by Arab and Armenian sources.

Conclusion

The question whether the military manuals and, by extension, the *ST* had any practical use on the battlefield is one that more or less impinges upon modern perceptions of what a practical manual looks like. Since modern manuals do not include antiquarian material and terminology as well as elements of promotional material and political propaganda, we may easily argue that any treatise which contains these was more of a literary project than a practical handbook. This is by no means a foregone conclusion though. We should take into consideration that Byzantine manuals were written for a literary audience which was very deeply influenced by tradition and most of the time viewed the past as a credible source to find exemplars or to be educated by it. Be that as it may, the presentation of the material in a proper way, that is, a way which was in line with traditional standards and connected to the past, would have been of vital importance, almost equally important as the contents of the work itself. Consequently, the work would look as if it was a gradual evolution of tradition, rather than a rigid breakup with traditional practices.

Undoubtedly, there could have been multiple purposes for the production of a manual. One might be the preservation of knowledge, another might be for an individual to promote himself against other peers, or for an emperor who was an armchair general to exercise propaganda referring to his predecessors or demonstrating that he had a more active role in the army. However, we cannot deny that practicality was among these reasons. Although a great deal of information came from older or classical works, the majority of them remained relevant to the tenth century or were updated to an extent. What is more, one can observe a gradual evolution of tactics from manual to manual, a fact which cannot be perceived as a coincidence. More importantly, there is good evidence from historical narratives, legal and administrative sources to support that a large percentage of the advice of the manuals was indeed followed on the battlefield or that Byzantine generals faced similar challenges. Last but not least, it would be futile to try to understand the practical value of military manuals without first trying to understand the role that tradition, history and the past played in the mentality of the Byzantine emperor, of the patrons and authors and of the military aristocracy.

Notes

1 For general remarks on the issue, see: Kolias 1997: 153–64; Cheynet 2014: 55–6; Theotokis 2014: 106–18; Whately 2015: 249–61; Rance 2017a: 292–6. For the practical use of *Apparatus Bellicus* see Zuckerman 1994: 385–9.
2 Sullivan 2010a: 160.
3 Hunger 1978: ii.323–4 (my trans. from the German); Dagron and Mihăescu 1986: 139–41 (my trans. from the French); McGeer 1995a: 171; Pryor and Jeffreys 2006: 1–6, 445–53; Holmes 2010: 61–80.
4 Dain and Foucault 1967: 350–1; Hunger 1978: ii.333; Krentz and Wheeler 1994: xxi–xxiii.
5 See Chapter 2, pp. 27–9.
6 See Chapter 5, pp. 72–3.
7 Cutler 1995: 203–5; *ODB:* ii.997–8; Dagron and Mihăescu 1986: 41, n.7; c.f. Spanos 2010: 51–9.
8 Berger 2013: 247–58; Mango 1975: 6–17; Shepard 1995: 108; Sinclair 2012: 202–3; Strässle 2006: 55–6. See also Kaldellis 2015b: 22–35.
9 Németh 2013: 232–58, demonstrates how the *Excerpta* of Constantine VII included works on stratagems and warfare from older authors and how these were popular with the Byzantine aristocracy. Kaegi 1990, argues in favour of the educational use of history for Byzantine officers. Alexopoulos 2008–2012: 47–71, argues that Alexius I used ancient stratagems to fight his battles and Neville 2012: 29–45, explains the use of ancient figures as exemplars in historiography.
10 *ST,* 1.10, 14.4 (trans. Chatzelis and Harris, pp. 22, 33), with Basil I, *Hortatory Chapters,* 1.65.
11 Psellos, *Chronographia,* 7.180; Anna Komnene, 15.3.6.
12 Constantine VII, *Three Treatises,* C.196–9; Theophylact Simocatta, 1.14.2 (trans. Whitby and Whitby, p. 40).
13 Julian, 3.124a-d; Libanius, 18.53,72,233.
14 Ammianus Marcellinus, 24.2.14–17; Lendon 2005: 293–309. See also Kaegi 1981; 1964b who further speculates about the works known to Julian.
15 *LT,* 15.29.
16 *ST,* 1.24 (trans. Chatzelis and Harris, pp. 24–5).
17 Kekaumenos, 2.
18 Urbicius, 2; 15 (trans. Burges and Greatrex, pp. 60, 67).
19 *MS,* introduction, 14–21 (trans. Dennis, p. 8). See also: Rance 1994: 85.
20 *LT,* 19.1.
21 Constantine VII, *Three Treatises,* C.20–39.
22 Constantine VII, *Three Treatises,* B; Haldon 1990a: 45.
23 *Geoponika,* 1.14.11.
24 *ST,* 1.1.
25 *ST,* 47.1, 20. See also Chapter 7, pp. 158–60.
26 *ST,* 30, 31; See McGeer 1995a: 182 who states that the Macedonian phalanx was the standard exemplar for the infantry in the tenth century.
27 *ST,* 38, 39. Maurice and Arrian tried to justify their innovations in the same way, see: Rance 1994: 111, 119.
28 *PM,* 1.63–74; McGeer 1995a: 182, 187; Sinclair 2012: 202.
29 *TNO,* 65.139–47, 169–72; Dain 1937: 13; Dain and Foucault 1967: 371; McGeer 1991: 137–8; c.f. Holmes 2010: 77 who refers to another similar example from the *De Obsidione,* where the author proposes the use of siege engines as the most effective way.
30 See for example, *DV,* preface, 19.5; 21.1; Sullivan 2010a: 157–8.
31 *DV,* 16.8 (trans. Dennis, p. 203).
32 Kekaumenos, 2 [trans. Roueché (online publication)].
33 Kaegi 1983: 20–1.

34 *De Obsidione*, 76–7; Sullivan 2000: 4–15; 2003a: 144–5.
35 See for example, Aelian, 25.1–2,9, 31.1,3, 36.1–2, 38.1, 39.1; *PS,* 22–3, 31; Rance 1994: 76–7; 2007a: 703–4; Zuckerman 1990: 217–19; Haldon 2014: 40; Dain 1938: 9; 1937: 87–9.
36 Holmes 2010: 61–77.
37 *LT,* 9.14, 18.95.
38 Holmes 2010: 66; Magdalino 2013: 189, 195; Tougher 1997: 166–8.
39 Dagron and Mihăescu 1986: 160–5; Dennis 1985: 139.
40 *DV,* preface; Dennis 1985: 149; Dagron and Mihăescu 1986: 33–7; Holmes 2010: 74.
41 Holmes 2010: 74–5; Cheynet 1986a: 312.
42 Pryor and Jeffreys 2006: 183–5, 522; Holmes 2010: 71–2; Angelidi 2013: 16–17.
43 Pryor and Jeffreys 2006: 185.
44 Pryor and Jeffreys 2006: 4–5, 183–6.
45 *LT,* 11.21; 15.32; 17.65.
46 Holmes 2010: 75; Dagron and Mihăescu 1986: 163–4, 171–5; Dennis 1985: 140.
47 Magdalino 2013: 189–209; Holmes 2010.
48 Dennis 1981: 16–18; 1984: xv–xvii; Dain and Foucault 1967: 344–6; Rance 1994: 36–42.
49 Dain 1937: 47–51; Dain and Foucault 1967: 370–1; Dagron and Mihăescu 1986: 154; Kolias 1993a: 23–4.
50 McGeer 1995a: 313–15.
51 *PS*, 13.69–71; 19.40–65; Dennis 1985: 3; Sullivan 2010a: 151–2; Dain and Foucault 1967: 343; Hunger 1978: ii.327–8; Rance 1994: 58–60; 2007a: 706; Zuckerman 1990: 215; Cosentino 2000: 277–80.
52 Rance 2007a: 719–37; Cosentino 2000: 262–80.
53 Holmes 2010: 61–2, 72–3.
54 McGeer 1991: 129–31; Dain 1937: 136; Trombley 1997: 269.
55 Skylitzes, 341–2; Dain 1937: 144; Dain and Foucault 1967: 373; Hunger 1978: ii.323, 336–7; McGeer 1991: 130–1.
56 *ST*, 1.1.
57 *MS*, intro.10–14, 22–31 (trans. Dennis, pp. 8–9); Dennis 1981: 13–16; Hunger 1978: ii.329–30; Kolias 1993a: 39–44. For the use of simple language and the military 'slang' in the *MS* see: Rance 2004a: 265–326; 2004b: 96–130; Gyftopoulou 2013: 84–6.
58 *LT*, epologue.71; Grosdidier de Matons 1973: 229–31; Tougher 1997: 169–71; Haldon 2014: 22–5.
59 Heron of Byzantium, *Parangelmata Poliorketika*, 1–4, 58; Sullivan 2000: 15–21; c.f. Holmes 2001: 480, who states that the author fails to keep a simple style and argues that the manual was more a court debate on how to use the past for the current needs, rather than something used by generals. See also Sullivan 2010a: 155.
60 *DV*, preface.1 (trans. Dennis, p. 147).
61 Pryor and Jeffreys 2006: 599.
62 Constantine VII, *Three Treatises*, C.196–9.
63 For the manual see: Kennedy 2001: 111–14; Haldon 2014: 43–4; Shatzmiller 1992: 253–7.
64 Sarraf 2004: 178–9, 195, 198.
65 Christides 1984: 137–43; 1995: 83–96; Serikoff 1992: 57–61. The texts of the Arab translations have been edited and translated by Ahmad Shboul in Pryor and Jeffreys 2006: 645–66.
66 Christides 1984: 139; Serikoff 1992: 58.
67 Pryor and Jeffreys 2006: 656–7, 660.
68 See for example: Strässle 2006: 51, 55–8; Kaegi 1983; Rance 1994: 142–7; Dennis 1997: 178; Haldon 1999: 149–50, 172; Sullivan 2003b: 520–1; Luttwak 2009: 235–408; Sinclair 2012: 201.

69 See Chapter 5 and Chapter 6 for more detail.
70 Riedel 2010; Riedel 2018: 32–73, 174–87; Kolia-Dermitzaki 1989: 41–9.
71 Featherstone 2013: 139–44.
72 See Chapter 5, pp. 78–83.
73 Dennis 1984: vii–xv; Dain and Foucault 1967: 344–5; Hunger 1978: ii.329–30; Rance 2005: 429–31, 435–6, 450, 455, 460–1, 464, 468–9; Syvänne 2004: 16–19.
74 Kaegi 1992: 57–61, 100–11, 123–8, 251–6; 2003: 117, 129–30, 161–3, 168; Trombley 2002: 241–52, 258–9; Karantabias 2005–2006: 28–41; Whitby 1988: 130–2, 138–83, 276–304.
75 Kolias 1984a: 129–35; Dagron 1983: 219–43; Haldon 2014: 33–8; Riedel 2010.
76 McGeer 1991: 129–38; 1995a: 174–360; Haldon and Kennedy 1980: 97, 105; Dagron and Mihăescu 1986: 215–87; Dennis 1985: 137–41; Haldon 2014: 100–1; Theotokis 2012: 5–15; Decker 2013: 132–40.
77 Sullivan 2000: 5–21; 2003a: 139–45.
78 *VB*, 46 (trans. Ševčenko, p. 165); Skylitzes, 141
79 Sinclair 2012: 267–72; Holmes 2005: 202–16; Markopoulos 2009: 702.
80 Sinclair 2012: 171–4, 197–208. See also: Mullett 1997: 41–3; 1981: 82, 88–9; Dennis 1988: 161; Ševčenko 1992: 189–93; Shepard 1995: 115; 2003b: 109–13.
81 McGeer 1995a: 191–5; Sinclair 2012: 201.
82 Psellos, *Chronographia*, 7.137 (trans. Sewter, p. 353), see also 1.33.
83 *VB*, 36 (trans. Ševčenko, p. 135).
84 Bryennios, 1.1.
85 The *VB* is largely promotional, and Psellos usually records Basil II ideally and Romanos IV badly, see: Krallis 2012: 81–100 and the discussion later.
86 Skylitzes, 245–6; LD, 7.
87 Talbot and Sullivan 2005: 30; Makrypoulias 2000: 355–6.
88 Other examples can be found in Skylitzes, 7, 32.
89 Skylitzes, 45–6; TC (b), 2.25; Genesios, 2.12. For the *thema* of the Kibyrrhaiote see: *TIB:* viii. 79, 116–25, 407–13.
90 Tabarī, 2103; Vasiliev 1935–1968: ii.102–3.
91 Skylitzes, 141–2; *VB*, 51.
92 See the discussion earlier and later for other parallel passages. For the *Alexia*d, see: Sullivan 2010b: 51–6.
93 See for example: Riedel 2012: 578–83; Sinclair 2012; Kaldellis 2012: 206; Neville 2012: 29–45; Lilie 2014: 157–210.
94 Holmes 2005: 202–39, 278–89; Talbot and Sullivan 2005: 13–14; Kaldellis 2013: 45–8; McGrath 1995: 152–64; Howard-Johnston 1983: 239–46; 2000: 302–3; McCormick 1986: 192–4; Sinclair 2012: 150–1, 153–7. For the authority of the aristocracy and its relationship with its peers and Constantinople see: Andriollo 2017.
95 TC, 427–8; Skylitzes, 230; Psellos, *Historia*, 98; Markopoulos 2006; c.f. Treadgold 2013: 198 who argues that Manuel's work was not a biography of Kourkouas, but a wider history in which Kourkouas was featured.
96 For lost promotional literature in Theophanes Continuatus, Leo the Deacon, Skylitzes and *Vita Basilii*, see among others: Sinclair 2012: 267–316; Markopoulos 2003: 186–8, 193–6; 2006; 2009: 697–710; Morris 1988: 85–6; Ljubarskij 1993: 252–3; Roueché 1988: 127–8; McGrath 1995: 152–64; Karpozilos 2002–2009: ii.352–7, 358–64, 399–400, 480–5; iii.248–9; van Hoof 2002: 163–83; Holmes 2005: 268–98; Kazhdan 2006: 167, 273–4, 279–80; Kiapidou 2010; Kaldellis 2013: 45–8; Treadgold 2013: 226–70, 329–39; Andriollo 2014: 119–31.
97 Sinclair 2012: 158–89, 390–7; Magdalino 1993: 314; Shepard 2005: 178–80; McCormick 1986: 42–4, 190–6; Walker 1977: 301–5, 319–27; Kaldellis 2014: 235–40; Howard-Johnston 2010: 422–23; 1994: 70–1; Whitby 1988: 97; Kaegi 2003: 131–2, 156.

98   For an overview see Sinclair 2012: 302–18.
99   For two different perspectives see: Holmes 2005: 284–7 and c.f. McGeer 1995a: 294–9.
100 Sinclair 2012: 306–8, 361–5; Dagron and Mihăescu 1986: 165–70; Cheynet 1986a: 293–6, 299–301.
101 For the ideal general see a forthcoming paper untilled 'The ideal general and the impact of Onasander and rhetoric on Middle Byzantine military manuals' by the present author. There is also a forthcoming paper by Philip Rance entitled 'The ideal of Roman general in Byzantium: The reception of Onasander's *Strategikos* in Byzantine military literature'. For the ideal Byzantine general from a Christian point of view see: Riedel 2018: 74–94.
102 Strässle 2006: 52; Haldon 1999: 37.
103 Lilie 2014: 168–76.
104 *ST*, 1.7–9 (trans. Chatzelis and Harris, p. 22)
105 Psellos, *Chronographia*, 1.32 (trans. Sewter, p. 46).
106 LD, 9–10 (trans. Talbot and Sullivan, pp. 62–4); Hoffmann 2007: 113–15.
107 *ST*, 1.10, 1.20, 1.23 (trans. Chatzelis and Harris, pp. 22, 24).
108 LD, 66–7 (trans. Talbot and Sullivan, pp. 115–17); Skylitzes, 267.
109 Bikhazi 1981: 846.
110 Lilie 2014: 168–76.
111 *ST*, 5 (trans. Chatzelis and Harris, p. 28); Onasander, 42.2; *MS*, 8.1.1.
112 LD, 75; *VB*, 40.7–15. The event is also described in *LT*, 9.14. For Antioch see *TIB:* v. and xv.
113 See for instance: Jenkins 1954: 20–30; van Hoof 2002: 163–83; Mango 2011: 4, 10–11; Karpozilos 2002–2009: ii.331–8, 352–6; Kazhdan 2006: 137–44; Treadgold 2013: 165–76; Markopoulos 2009: 697–706. For the campaigns of Basil, I against the Arabs see: Tobias 2007.
114 Vasiliev 1935–1968: ii.i.331–2; Canard 1951: 779.
115 *ST*, 1.21–2, 6 (trans. Chatzelis and Harris, pp. 24, 29); Onasander, 3.1, 9.24.
116 Psellos, *Chronographia*, 1.28 (trans. Sewter p. 43).
117 Holmes 2005: 282–4.
118 Tabarī, 2239 (trans. Rosenthal, pp. xxxviii. 136).
119 See the discussion on acquiring information later.
120 *ST*, 4.1–3 (trans. Chatzelis and Harris, p. 28); Onasander, 33.
121 *MS*, 7.B1.8–10; *L.T.* 14.3, 14.99; *ST*. 46.17; McGeer 1995a: 285.
122 *PM*, 4.121–53; McGeer 1995a: 307–8; Dennis 1997: 174–5; Haldon 1999: 228–31.
123 Talbot and Sullivan 2005: 13–14; Morris 1994: 209; McGrath 1995: 156–9; Holmes 2005: 240–98; Sinclair 2012: 319–85; Kaldellis 2013: 35–52; Neville 2012: 2–27, 121–38, 194–203; Markopoulos 2009: 697–714; Andriollo 2014: 126–38; Lilie 2014: 188–90; Kazhdan 2006: 273–8.
124 LD, 29–30, 96–7, 110–11 (trans. Talbot and Sullivan, pp. 82, 146); Skylitzes 290–1. See Holmes 2005: 272–6, for the differences in the accounts of Leo the Deacon and Skylitzes.
125 TC, 404–5.
126 Symeon Magister, 136.27; Skylitzes 218; Holmes 2005: 150–1; Sinclair 2012: 355.
127 However, in 953, the Byzantine commander Bardas Phokas was wounded in the face, which might imply he had a more active role in battle, see: Skylitzes, 241; Vasiliev 1935–1968: ii.i.323–6, 362; ii.348–51; Canard 1951: 774–6; Bikhazi 1981: 747–8.
128 Vasiliev 1935–1968: ii.i.121–2.
129 Sinclair 2012: 361–5.
130 For some considerations of the ideal general in the treatises see Strässle 2006: 371–6 and Chapter 5.
131 *MS*, 7.6,15, 8.1.2, 3, 15; *LT*, 13.6, 19.20, 20.18.

132 *ST*, 17.1–2 (trans. Chatzelis and Harris, p. 35).

133 Attaleiates, 20.153; Skylitzes, 17 (trans. Wortley, p. 19); Genesios, 1.15; TC (b), 1.14.

134 LD, 57–8. For the legal punishment of throwing the equipment during the battle see: Ashburner 1926: 93; McGeer 1995a: 335–8.

135 Psellos, *Chronographia*, 1.33 (trans. Sewter, p. 47).

136 TC (b), 2.19; Genesios, 2.8; Skylitzes 40. For Bizye see *TIB:* xii. and vi.

137 Skylitzes, 217 (trans. Wortley, p. 211); Symeon Magister, 136.26; TC, 404.

138 *ST*, 15.1 (trans. Chatzelis and Harris, p. 34).

139 Skylitzes, 68 (trans. Wortley, p. 69); TC (b), 3.22.

140 Vasiliev 1935–1968: i.301; Tabarī, 1243. For Amorion see: *TIB:* vii. and iv.

141 *ST*, 50.2–3 and 1.26 where the original source is Onasander, 34.1–2 (trans. Chatzelis and Harris, p. 84). See also Constantine VII, *Three Treatises*, B.92–4, where the emperor is reported to dine with two or three table guests, and *LT*, 16.1–5.

142 Skylitzes, 68 (trans. Wortley, p. 69); TC(b), 3.22; Grégoire 1934: 183–204; c.f. Genesios, 3.9; Treadgold 1979: 180–2 who argues that Manuel saved the emperor. For the different accounts see: Codoñer 2014a: 132–5.

143 LD, 53 (trans. Talbot and Sullivan, p. 103); c.f. Skylitzes 269 who confirms the reward of the army, but states that they were given a specified amount of the booty. Leo the Deacon dates the fall of the city in 964, but Skylitzes, 268 and Yahya, i.796, place it in 965. For full analysis see: Apostolopoulou 1982: 157–67 and for Mopsuestia, *TIB:* v.

144 LD, 159 (trans. Talbot and Sullivan, p. 201). For donatives and cash rewards see: Haldon 1984: 307–18.

145 *JGR*, i.247–8; McGeer 2000: 86–9.

146 Skylitzes, 250.

147 LD, 141, 159 (trans. Talbot and Sullivan, p. 186, 201). The statement 'made them even more zealous for battle' is suspiciously close to the advice of the *ST*, 50.2 and *LT*, 16.3, about how the spirits of the rewarded men will become more eager after their reward and could perhaps have been part of a lost promotional source, written by somebody who was aware of relevant passages from the manuals. For Dorostolon see: *TIB:* vi. 64, 88f., 91f., 95, 97f., 111, 131.

148 Psellos, *Chronographia*, 1.25; Holmes 2005: 285–91, 294–7.

149 *VB*, 12; TC, 403; Symeon Magister, 136.23; Skylitzes, 216–17 (trans. Wortley, p. 210).

150 Skylitzes, 26 (trans. Wortley, p. 29); TC (b), 2.5.

151 According to Mazzucchi 1978: 296–8; McGeer 2003: 116; c.f. Ahrweiler 1967: 402 who proposed the date of 952–3.

152 Constantine VII, *Military Oration*, 85–93 (trans. McGeer, p. 120).

153 *ST*, 50.4 (trans. Chatzelis and Harris, p. 84).

154 LD, 23, 27, 32, 53 (trans. Talbot and Sullivan, pp. 75, 79, 84); Skylitzes, 254, TC, 306.

155 *Procheiros Nomos*, 40; Schminck 1986: 98–101; c.f. van Bochove 1996: 29–56.

156 *LT*, 20.192; Kyriakidis 2009: 168; Dagron and Mihăescu 1986: 231–4; Kolias 1995: 131–2.

157 (trans. Chatzelis and Harris, p. 84).

158 Treadgold 1992: 91–2.

159 TC, 417–18, 429–30, 446–52; Symeon Magister, 136.57–59; Skylitzes, 225, 231. These expenditures seem to imply a sufficient economic stability during the reign of Romanos I which could have been used for army expenditures. This evidence can be seen together with the conversion of Melitene to a *kouratoria* (934), which provided additional money to the treasury of Romanos. TC, 443, speaks of Romanos' excessive demands from the poor of some *themata* which, if true, could mean that extra money was given to the army. This reference, however, could merely have been a rhetorical scheme to underline the injustice that supposedly existed in the times of Romanos I so as to portray Constantine VII as a better leader.

160 See Haldon 2013: 373–86 for a study of information gathering and guerrilla warfare.
161 *ST*, 8.2; 44.4; 51.1. The sources of these passages can be traced to *MS*, 8.1.32, 10.2.23–29, 7.13B; *LT*, 14.25.
162 Skylitzes, 241; Vasiliev 1935–1968: ii.i.323–6; ii.348–51; Canard 1951: 774–6; Bikhazi 1981: 747–8.
163 Canard 1951: 812; Bikhazi 1981: 856–61.
164 *ST*, 7.2 (trans. Chatzelis and Harris, p. 29). See also 44.3.
165 *MS*, 7A.13–20, 7.3.
166 LD, 109, 120 (trans. Talbot and Sullivan, pp. 159, 168).
167 LD, 108.
168 LD, 9, 19.
169 Ljubarskij 1993: 252–3; Talbot and Sullivan 2005: 14; Morris 1988: 85–6; McGrath 1995: 153–6, 162; Kaldellis 2013: 42; Sinclair 2012: 50–7, 361–5; Holmes 2005: 273–6, 278–89; Kazhdan 2006: 283; Riedel 2012: 580–3.
170 TC, 475–6; Kaldellis 2015a: 302–11.
171 Symeon Magister, 136.17; Vasiliev 1935–1968: ii.i.189, ii.259.
172 *DC*, 657.19. For further evidence on intelligence and its capabilities in the tenth century see: Shepard 1995: 109–16.
173 For the use of espionage by the Byzantines see: Koutrakou 1995: 135–44.
174 *PS*, 42; *MS*, 8.2.26, 9.5.51–55; *LT*, 20.84; *ST*, 25.1 (trans. Chatzelis and Harris, p. 44).
175 See for example, Beihammer 2012: 387–400; Haldon 1999: 224; Decker 2013: 144–8.
176 LD, 108; Tabarī, 1246; Vasiliev 1935–1968: ii.i.302.
177 *ST*, 25.2 (trans. Chatzelis and Harris, p. 44).
178 *DV*, 7.2.
179 Vasiliev 1935–1968: ii.i.415–16.
180 *ST*, 27.2, 44.7 (trans. Chatzelis and Harris, pp. 45–6). These passages seem to derive from *MS*, 8.1.36, 9.3.21–31; *LT*, 17.32,92 20.38.
181 Vasiliev 1935–1968: ii.268; Canard 1951: 735; Bikhazi 1981: 422–5.
182 Skylitzes, 242; Vasiliev 1935–1968: ii.341–5; Koutrakou 1995: 142.
183 Leo the Deacon is also silent about the capture of Cyprus by Niketas in 965. For the capture of Cyprus see Sinclair 2012: 56; Lemerle 1972: 153–4; Savvides 1993: 371–8.
184 Vasiliev 1935–1968: i.137–41, 293–4; Tabarī, 1235–6. For Zapetra see: *TIB:* ii.286f. and v.49.
185 *ST*, 44.7 (trans. Chatzelis and Harris, p. 62).
186 LD, 14. See Koutrakou 1995: 123–32, for more unusual means of gathering intelligence.
187 Vasiliev 1935–1968: i.296–7; Tabarī, 1238–9. For a discussion about Mutasim's march through Cappadocia during the expedition of Amorion see: Vasiliev 1935–1968: i.149–76; Bury 1909: 120–9.
188 *ST*, 27.2 (trans. Chatzelis and Harris, pp. 45–6).
189 Vasiliev 1935–1968: i.298–300; Tabarī, 1241–2.
190 Vasiliev 1935–1968: i.220, 366–7.
191 Onasander, 39.4; *LT*, 15.37; *ST*, 11.5 (trans. Chatzelis and Harris, p. 32); Sullivan 1997: 193.
192 Yahya, i.74; Vasiliev 1935–1968: ii.i.97, 125, 337–40; ii.335; Canard 1951: 781; Bikhazi 1981: 778–9.
193 Kaminiates, 29. Some scholars have questioned the narrative of Kaminiates, see for example, Kazhdan 1978: 301–14; Treadgold 2013: 121–3, but others have defended his credibility, see for example, Karpozilos 2002–2009: ii.270–3.
194 *MS*, 8.1.10; *LT*, 20.13; *ST*, 7.1 (trans. Chatzelis and Harris, p. 29).
195 Constantine VII, *Military Oration*, 48–51 (trans. McGeer, p. 119).
196 Symeon Magister, 131.2–4; Vasiliev 1935–1968: ii.i.195–6.

197 Canard 1951: 782.
198 Bikhazi 1981: 612–13. For earlier examples see: Rance 1994: 202–7. Disinformation was regularly employed by Byzantine diplomacy; for an analysis with a focus on diplomatic relations with the West see: Shepard 1985: 275–92.
199 *ST*, 22.4 (trans. Chatzelis and Harris, p. 39); Onasander, 8,9; *MS*, 4.3.53–56, 5.1, 7.13; 12.22; *LT*, 11.3, 8, 13–18, 30–3; *PS*, 29.25–30.
200 LD, 142–3 (trans. Talbot and Sullivan, p. 187). The manuals record the use of the spear wall in our period, see: *TNO*, 62.31–5 and something similar in *PS*, 27; McGeer 1995a: 348–54.
201 LD, 16–17, 58–9 (trans. Talbot and Sullivan, pp. 68, 106).
202 *ST*, 22.3 (trans. Chatzelis and Harris, p. 39).
203 LD, 133, 171–2 (trans. Talbot and Sullivan, pp. 179, 214). On the other hand, there is also the more neutral account of TC (b), 2.18; Skylitzes, 38 who records the choice of a camp which was in suitable location because of the flowing rivers around it during the reign of Michael II. For Preslav see: *TIB:* vi.50, 88, 92, 94, 96, 111, 131, 143, 202, 319, 339, 450, 457.
204 Talbot and Sullivan 2005: 11; Sinclair 2012: 57–60; Karpozilos 2002–2009: ii.492–501; Treadgold 2013: 241–4.
205 Skylitzes, 249 (trans. Wortley, p. 240). Skylitzes, 300 also agrees that Tzimiskes had 'established a well-fortified camp' before Dorostolon, but this belongs to his generalized phrases.
206 Holmes 2005: 94–5; Morris 1988: 85–8; Sinclair 2012: 61–3; Kiapidou 2010: 353–9.
207 Holmes 2005: 150–1.
208 TC, 476; Kaldellis 2015a: 302–11.
209 *VB*, 51 (trans. Ševčenko, pp. 185–7).
210 Vasiliev 1935–1968: ii.i.368; Yahya, ii.497, 525, which is also confirmed by Attaleiates, 17.109. For speculations concerning the use of Greek sources by Yahya see Forsyth 1977: 182–98, who concluded that Yahya relied on Greek sources more heavily for the first part of the tenth century but claims that the only extant source which can be identified with some security is probably Symeon Logothetes.
211 Tabarī, 2030; *ST*, 21, 22.
212 *ST*, 22.5; *MS*, 4.3.53–56; *PS*, 29.25–28; *LT*, 11.8. For the camp in *MS* see Gyftopoulou 2013: 77–80.
213 Anna Komnene, 5.4.5.
214 William of Apulia, 4.11–13. For the debate on William of Apulia, his sources, and the relationship of his work with Anna see: Mathieu 1961: 38–46; Loud 1991: 48–55; Brown 2011: 162–73; Frankopan 2013: 80–99.
215 Tsurtsumia 2011a: 415–21.
216 Haldon 1999: 210–12.
217 *ST*, 22.11 (trans. Chatzelis and Harris, p. 41).
218 Theophanes Confessor, AM6200.
219 Skylitzes 269, and for a similar case, 302.
220 Yahya, i.79.
221 *VB*, 51 (trans. Ševčenko, p. 187).
222 *ST*, 22.7–8, 12, 49.9.
223 See pp. 100–1 above.
224 LD, 9–10.
225 Hoffmann 2007: 113–15.
226 TC, 476–7; Kaldellis 2015a: 306–7.
227 LD, 171–2 (trans. Talbot and Sullivan, p. 213).
228 Holmes 2005: 226–8.
229 LD, 143–4; c.f. Skylitzes, 300 (trans. Wortley, p. 286).

230 Symeon Magister, 135.23; TC, 390; c.f. Skylitzes 204–5. For a discussion of the battle and the different narratives in the sources see Grigoriou-Ioannidou 1983: 123–48.

231 The chapter which covers night attacks is found in *ST*, 48. Similar material is found in *MS*, 8.1.25, 9.2; *LT*, 17.14–19; *PS*, 39.39–43. For night attack credentials see *ST*, 48.8 and Chapter 5, pp. 76–7.

232 LD, 19–20 (trans. Talbot and Sullivan, p. 72). For the pro-Phokas bias see: Talbot and Sullivan 2005: 14; Morris 1988: 85–6; Sinclair 2012: 50–6; Kazhdan 2006: 283. Leo Phokas was the subject of many promotional accounts; for his skill in surprise attacks see: Dennis 1985: 139–40; Cheynet 1986a: 302–5. For a detailed discussion of this battle and of the conflicts between the Byzantines and Magyars see: Grigoriou-Ioannidou 1999: 65–135.

233 Skylitzes, 67 (trans. Wortley, p. 69); TC (b), 3.22.

234 Genesios, 3.8.

235 Codoñer 2014a: 132–5; Kaldellis 1998: 45–7.

236 *DC*, 651–2, 664–78; Tsougarakis 1988: 61–4.

237 LD, 14.

238 *LT*, 18.121 (trans. Dennis, p. 483).

239 *PM*, 4.191–212 (trans. McGeer, p. 41).

240 *ST*, 48.1–2, 48.6 (trans. Chatzelis and Harris, p. 80).

241 *VB*, 42; LD, 14. For the Armeniakon *thema* see: *TIB*: ix.42, 46, 69f., 73–7, 81–6, A. 209, 141.

242 LD, 14 (trans. Talbot and Sullivan, p. 67); *VB*, 42 (trans. Ševčenko, p. 153). Genesios, 4.36, has a similar narrative with the *VB*, because he seems to have used the latter as a source, see for example, Treadgold 2013: 186–8. LD, 18–19, also has Leo Phokas attack the enemy camp by dividing his army into three sections. Given the lack of any further detail, this could either denote the standardized phrase used by historians to describe a drawn-up formation or adherence to the advice of the manuals not to surround the enemy camp completely.

243 *ST*, 48.2; LD, 14.

244 *ST*, 1.18 (trans. Chatzelis and Harris, p. 24).

245 Skylitzes, 341–2. For Sperchios see *TIB*: i.

246 TC, 462–3.

247 Much of the advice of the *ST* concerning siege warfare derives from: *MS*, 10.3; *LT*, 15; *PS*, 13. For a general discussion of siege warfare in Byzantium see: McGeer 1995b; Sullivan 1997; 2003a, 2010b; Haldon 1999: 134–8, 181–9; Decker 2013: 158–61; Petersen 2013: 115–43.

248 *ST*, 53.1, 53.4 (trans. Chatzelis and Harris, pp. 86–7); *De Obsidione*, 57.71, 59.84. For food supplies and whether this advice was followed see also Sullivan 2003b: 511–12.

249 *ST*, 53.4 (trans. Chatzelis and Harris, p. 87).

250 *ST*, 53.1 (trans. Chatzelis and Harris, p. 86).

251 Vasiliev 1935–1968: ii.i.79, 158–60, ii.366–7.

252 Skylitzes, 39 (trans. Wortley, p. 42); TC (b) 2.19.

253 Theophanes Confessor, AM6206, 6208 (trans. Mango and Scott, p. 540); Yahya, ii.497.

254 *LT*, 20.81; Kaminiates, 21–3; Kolias 1989: 467–76; Makrypoulias 2012: 109–20.

255 *ST*, 53.4 (trans. Chatzelis and Harris, p. 87).

256 LD, 58 (trans. Talbot and Sullivan, p. 106).

257 Bikhazi 1981: 854–5.

258 *ST*. 53.1 (trans. Chatzelis and Harris, p. 86).

259 Skylitzes, 300 (trans. Wortley, p. 286). For Dristra see Dorostolon, *TIB*: vi.64, 88f., 91f., 97f., 111, 131.

260 Skylitzes, 78; TC (b), 3.34. Arab sources speak of two acts of treason see: Tabarī, 1245–6, 1251–2; Vasiliev 1935–1968: i.301–3.

261 Vasiliev 1935–1968: ii.346; Sullivan 1997: 191.
262 Skylitzes, 272–3; LD, 81–2; Yahya, i.99–100, 108.
263 Kekaumenos, 2; Haldon 1999: 186–7.
264 Vasiliev 1935–1968: ii.264; Canard 1951: 733.
265 *ST*, 53.1–2.
266 Tabarī, 1250; Kaminiates, 27; Vasiliev 1935–1968: i.306. For other examples see: Sullivan 2003b: 517–18.
267 Tabarī, 1251–2; Vasiliev 1935–1968: i.305–6.
268 For the advice of other manuals and parallel narratives see also: Sullivan 2003b: 514–17.
269 *ST*, 53.5 (trans. Chatzelis and Harris, p. 87).
270 *ST*, 53.6 (trans. Chatzelis and Harris, pp. 87–8).
271 *ST*, 53.8 (trans. Chatzelis and Harris, p. 88).
272 *ST*, 53.7 (trans. Chatzelis and Harris, p. 88).
273 Tabarī, 1245, 1248; Vasiliev 1935–1968: i.302–4
274 Kennedy 2001: 184–5.
275 Kaminiates, 29 (trans. Frendo and Fotiou, pp. 51–3).
276 Tabarī, 1245; Anna Komnene, 7.8; Vasiliev 1935–1968: i.302
277 Albert of Aachen, 6.9.
278 *ST*, 53.5 (trans. Chatzelis and Harris, p. 87); *MS*, 10.3.9–13; *PS*, 13.71–81,115–20; *De Obsidione*, 69.
279 *DAI*, 51.114–20; McGeer 1991: 135–8; Sullivan 2000: 175; Kennedy 2001: 134.
280 Skylitzes, 463. For Manzikert see: *TIB*: ii. 84, 99, 103f.
281 See for instance: Korres 1995: 118–22; Haldon 1999: 189; 2000: 278–80.
282 Kaminiates, 33 (trans. Frendo and Fotiou, p. 59).
283 Tabarī, 1248; Albert of Aachen, 7.24; Vasiliev 1935–1968: i.302–6; Kennedy 2001: 134.
284 *DC*, 673.3–4.
285 Attaleiates, 8.46–47; c.f. Skylitzes, 463, who gives an alternative testimony. See also: Vasiliev 1935–1968: ii.i.150, ii.262–3. For Dvin see: *TIB*: ii.86.
286 *VB*, 59 (trans. Ševčenko, p. 213); Skylitzes 151; LD, 16 (trans. Talbot and Sullivan, p. 68); Sullivan 2000: 161. For Euripos see: *TIB*: i.
287 Kaminiates, 26 (trans. Frendo and Fotiou, p. 47).
288 Tabarī, 1248; Ralph of Caen, 160.
289 *ST*, 53.3, 53.10 (trans. Chatzelis and Harris, pp. 87–8).
290 LD, 148 (trans. Talbot and Sullivan, p. 192).
291 Attaleiates, 8.46–7; c.f. Skylitzes, 463, who gives an alternative testimony. The credibility of Attaleiates as an eyewitness account has most recently been challenged by Vratimos 2012: 829–40, see also: Krallis 2012: 126–56.
292 TC, 404–5; Symeon Magister, 136.27; Skylitzes 218. See also: Haldon 1999: 184–5.
293 Vasiliev 1935–1968: ii.i.97, 125, 337–40, ii.355. For additional examples see: Sullivan 2003b: 518.
294 Vasiliev 1935–1968: i.359–60, ii.i.146
295 *ST*, 53.9 (trans. Chatzelis and Harris, p. 88).
296 *ST*, 53.9.
297 *Hypothesis*, 56.7; *PS*, 13.34–43; Heron of Byzantium, *Parangelmata Poliorketika*, 11; *De Obsidione*, 76.185–95.
298 Polybius, 21.28. For Ambracia see: *TIB*: iii.104.
299 Anna Komnene, 13.3; Sullivan 2010a: 52. For Dyrrachium see: *TIB*: iii.
300 For Anna, her credibility and her sources, see: Neville 2016: 47–59, 75–88; Sinclair 2014: 143–85; Macrides 2000: 63–82; Magdalino 2000: 15–44; Buckley 2014; Frankopan 2012: 38–88; 2014: 38–52. For special focus on siege description, see: Sullivan 2010b: 51–6.

301 Theophanes Confessor, AM6268; Tabarī, 1247; Skylitzes, 302; Vasiliev 1935–1968: i.302–6.
302 See also: Sullivan 2003b: 520–1, who also argues in favour of the practical use of manuals regarding defensive siege warfare.
303 *ST*, 11.1; 12; Sullivan 1997: 186.
304 Haldon 1999: 183–4.
305 *ST*, 11.4 (trans. Chatzelis and Harris, p. 32).
306 For example, see: Vasiliev 1935–1968: ii.i.260; Canard 1951: 774–5.
307 LD, 16–17, 60, 124 (trans. Talbot and Sullivan, p. 108); Haldon 1999: 185.
308 Sinclair 2012: 361–2; Morris 1988: 85–6; Kazhdan 2006: 283; Riedel 2012: 580–3; Hoffmann 2007: 127–30.
309 Skylitzes, 271–2.
310 Holmes 2005: 150.
311 Skylitzes, 302–3 (trans. Wortley, p. 288); LD 135–6.
312 TC, 404–5; Symeon Magister, 136.27; Skylitzes 218; Vasiliev 1935–1968: ii.268; Runciman 1929: 88–9.
313 For example, see: LD, 60; Skylitzes, 231–2; *VB*, 38; Vasiliev 1935–1968: ii.111; Bikhazi 1981: 783–4; Canard 1951: 748–51, 761.
314 TC, 415–16; Skylitzes, 224–5; Symeon Magister 136.53; Vasiliev 1935–1968: ii.268–9; El Cheikh 2004: 168. The same is recorded for Samosata, see: Canard 1951: 742.
315 Bosworth 1992: 279; Vasiliev 1935–1968: ii.269, n.1. For Arab sources recording Nikephoros Phokas' strategy and motives see also El Cheikh 2004: 168–78; Takirtako-glou 2015: 57–114
316 *JGR*, i.213–14 (trans. McGeer, p. 60); McGeer 2000: 49–53.
317 *ST*, 54; *MS*, 9.3.75–81, 10.1.11–18, 31–51; *LT*, 15.4–5, 21, 27.
318 *ST*, 54.2 (trans. Chatzelis and Harris, p. 89); Sullivan 1997: 188.
319 *ST*, 54.2 (trans. Chatzelis and Harris, p. 87).
320 Liudprand of Cremona, 1.10–11; Haldon 1999: 38.
321 Kaminiates, 28 (trans. Frendo and Fotiou, p. 51).
322 Tabarī, 1251; Vasiliev 1935–1968: i.306; Skylitzes 77–8; TC (b), 3.34; Genesios 3.11.
323 *ST*, 53.5, 54.5.
324 *ST*, 54.3, 54.4.
325 For example, see: LD, 25, 52; TC, 385, 415–16; McGeer 1995a: 123–9; Sullivan 1997: 179–80, 191; Dennis 1998: 99–115; Haldon 1999: 188–98.
326 Heron of Byzantium, *Parangelmata Poliorketika*, 53.37; LD, 25; Sullivan 2000: 239. For a study of military terminology and technology regarding siege warfare: Makryp-oulias 2013: 31–44.
327 Dennis 1998: 107; Haldon 1999: 188–98; Chevedden 2000: 74–84. For the other types of stone-throwing siege engines see Haldon 2000: 273–5.
328 *LT*, 15.27; Heron of Byzantium, *Parangelmata Poliorketika*, 46; Dain 1933: 16, n.2; Sullivan 1997: 191; 2000: 161.
329 For instance, see: LD, 82, 135–6; Skylitzes, 296–7.
330 *Miracula Demetrii*, 203.25; Theodosius the Deacon, 327; Sullivan 1997: 191.
331 *ST*, 54.3 (trans. Chatzelis and Harris, p. 89); *LT*, 15.27; *TNO*, 65.139–47; Haldon 1999: 186–7; McGeer 1995a: 125–8.
332 Heron of Byzantium, *Parangelmata Poliorketika*, 2.1–5. 2.6; 39.33–40.1. Sullivan 2000: 159–60; 1997: 179–81, 183, 190–1, 196–7.
333 LD, 25 (trans. Talbot and Sullivan, p. 77).
334 LD, 25, 52–3 (Talbot and Sullivan, pp. 77–8); see also Haldon 1999: 186–7.
335 LD, 26; Agathias, 1.10; Talbot and Sullivan 2005: 9, 11, 13–15, 78, n.41; Sinclair 2012: 59; Karpozilos 2002–2009: ii.492–501; Treadgold 2013: 241, 243–4; Sinclair 2012: 57–9.
336 *DC*, 670.10–11; Sullivan 1997: 191; Haldon 1999: 188; Vasiliev 1935–1968: ii.i.150. See also Anna Komnene, 13.3.

337 *DC*, 670.11,16, 671.4–5; Sullivan 1997: 191. See also *DC*, 659.1 with Haldon 2000: 270 for the equipment of the Byzantine navy with regard to the expedition during the reign of Leo VI.
338 *ST*, 54.4 (trans. Chatzelis and Harris, p. 89).
339 *LT*, 15.27; Heron of Byzantium, *Parangelmata Poliorketika*, 15.
340 Attaleiates, 8.46–47; c.f. Skylitzes, 463, who gives an alternative testimony.
341 Kaminiates, 29; *ST*, 54.4.
342 Albert of Aachen, 6.18; *DC*, 673.
343 LD, 171 (trans. Talbot and Sullivan, p. 214). For Serdica see: *TIB*: vi.61, 125f., 132, 190, 397.
344 Vasiliev 1935–1968: ii.i.149–51.
345 Kaminiates, 34 (trans. Frendo and Fotiou, p. 61).
346 Ibid.
347 *ST*, 54.6 (trans. Chatzelis and Harris pp. 89–90).
348 LD, 26–7 (trans. Talbot and Sullivan, pp. 78–9).
349 For the attempt of Leo the Deacon to underline Nikephoros' military understanding as well as mercy see Riedel 2012: 582–3.
350 Treadgold 1992: 78–145; 1995; Haldon 2000: 201–352; Cheynet 1995: 319–35; Whittow 1996: 181–93.
351 *ST*, 47.2,18.
352 *DC*, 651–2, 664–7. For interpretations of the numbers see: Haldon 2000: 236–334; Treadgold 1995; 1992: 78–145; Makrypoulias 2000: 352–4.
353 LD, 132–3; Vasiliev 1935–1968: ii.i.154. For historical narratives and exaggerated numbers see: Haldon 1999: 101–6; 2014: 102–6.
354 *LT*, 18.136–49; *DRM*, 1.11, 4, 8; Haldon 2014: 103–5.
355 *ST*, 46.26.
356 *PM*, 4; *DV*, 17 (trans. Dennis, p. 205); McGeer 1995a: 280.
357 *DC*, 661. *DV*, 16.4 records a thematic army of 3,000 men. See also: Haldon 2000: 326; Vlysidou et al. 1998: 215–16; Oikonomides 1996: 105–25; Zuckerman 2014: 193–233. For the *thema* of Peloponnesus see: *TIB*: i.59, 61, 66–8.
358 *LT*, 18.136–49.
359 For instance, a late ninth-century Arab source counts the total of the Byzantine cavalry at 40,000 men, see Haldon 2014: 96, 105. For a detailed discussion about theoretical registries and active soldiers see: Haldon 2000: 316–34. See also: LD, 132–3 who speaks of 13,000 cavalry.
360 *ST*, 45.2.
361 *PM*, 1; *LT*, 4.64–76; McGeer 1988: 135–45; 1995a: 257–80.
362 *ST*, 49.5.
363 *ST*, 47.3.
364 LD, 24.
365 *LT*, 18.113–15 (trans. Dennis, pp. 479–81). Some scholars have interpreted them as hollow, see for instance, Haldon 2014: 360.
366 Anna Komnene, 15.3–7 (trans. Sewter and Frankopan, pp. 440, 450–1).
367 *ST*, 46.10–13; Bryennios, 4.4–8; Anna Komnene, 1.5. See also Chapter 5, p. 82.
368 LD, 59, 140 (trans. Talbot and Sullivan, pp. 107–8).
369 Skylitzes, 298–9 (trans. Wortley, pp. 284–5).
370 *ST*, 46.2–3, 9.
371 Buckley 2006: 106; 2014: 264–7; Birkenmeier 2002: 5.
372 *ST*, 46.10–11, 46.12–13; Bryennios, 4.4–7, 4.8; Anna Komnene, 1.5; Tobias 1979: 200–1; Neville 2012: 92–131; Buckley 2014: 107–67. For Kalobrye see: *TIB*: xii.421.
373 McGeer 1995a: 313–16.
374 *VB*, 46 (trans. Ševčenko, p. 165).

375 Raymond of Aguilers, 133–5; France 1994: 361–3; Smail 1995: 122–4; Theotokis 2013: 57–71; Bennett 2001: 1–18.
376 Tabarī, 2240; *ST*, 46.12–13.
377 See for example, Treadgold 2013: 236–46.
378 *ST,* 39.1–6; 46.4,6–7
379 Treadgold 2013: 240.
380 *ST*, 39.1–6; Vasiliev 1935–1968: ii.i.333.
381 Mas'udi, 11 (trans. Lunde and Stone, p. 22).
382 Skylitzes, 298–9 (trans. Wortley, pp. 284–5); LD, 139–40; McGeer 1995a: 300–1.
383 *ST,* 23.4 (trans. Chatzelis and Harris, p. 42).
384 *ST*, 23.6 (trans. Chatzelis and Harris, p. 43).
385 *ST*, 23.2–3. Similar advice is of course found in *MS*, 9.4.10–20 and the *PS,* 18.31–8.
386 Skylitzes, 348–9.
387 Holmes 2005: 198–9.
388 *VB*, 49; Skylitzes, 143; Vasiliev 1935–1968: ii.i.174–5, ii.346; Bikhazi 1981: 723–4. For Adata see: *TIB:* ii.77, 83, 88–91, 127, 167f., 234.
389 LD, 171–3; Stephen of Taron, 186–7. For the different account of Skylitzes, 330–1 and its purpose see Holmes 2005: 224–8.
390 Vasiliev 1935–1968: ii.260–1, 305–6, ii.i.149–50; Canard 1951: 758.
391 Vasiliev 1935–1968: ii.i.308–9, ii.349–50; Bikhazi 1981: 747. See also Howard-Johnston 1983: 242–5 for the movements of Sayf in 956, which perhaps come from an official Ḥamdānid bulletin and may also have been idealized. However, he proposes that the victory of Sayf in this case may have needed 'little or no embellishment'.
392 Vasiliev 1935–1968: ii.i.126–7, ii.349; Canard 1951: 774–6.
393 *ST*, 52.1–2 (trans. Chatzelis and Harris, p. 86); *MS*, 10.2.1–7; *LT*, 17.59.
394 See for example, Stouraitis 2009: 45–168; McMahon 2016: 22–33.
395 Yahya, i.74–5; Vasiliev 1935–1968: ii.i.97, ii.340; Canard 1951: 790; Runciman 1929: 143–4; Vasiliev 1935–1968: ii.289–90.
396 Skylitzes, 242 (trans. Wortley, pp. 233–4).
397 LD, 19–23 (trans. Talbot and Sullivan, pp. 71–3).
398 Hoffmann 2007: 116–20. For battle exhortations in Middle-Byzantine historiography see Kyriakidis 2016b: 19–36. For a study of practices aiming to boost the morale of the Byzantine army see Karapli 2010.
399 LD, 19–23 (trans. Talbot and Sullivan, pp. 71–3).
400 Holmes 2010: 74–5; Cheynet 1986a: 312; Dagron and Mihăescu 1986: 33–7.
401 Cheynet 1986a: 293–4, 304–5; Sinclair 2012: 306–7; Markopoulos 2006.
402 Yahya, i.70–1; Vasiliev 1935–1968: ii.i.95–6, 309–12, ii.344–5; Canard 1951: 766–8; Bikhazi 1981: 716–21.
403 Yahya, i.83–4; Bikhazi 1981: 846
404 *ST*, 52.2 (trans. Chatzelis and Harris, p. 86).
405 Vasiliev 1935–68, ii.i.313, ii.344–5.
406 *PS,* 40; *LT*, 14.49.
407 *ST*, 24.3–5.
408 McGeer 1995a: 298–9.
409 Skylitzes 289–90.
410 Holmes 2005: 284–6.
411 LD, 108–10; Holmes 2005: 273–6. For a discussion as to whether Skylitzes had used Leo the Deacon as a source see: Karpozilos 2002–2009: ii. 477–8; Kiapidou 2010: 104–10; Treadgold 2013: 239–40.
412 *ST*, 52.3 (trans. Chatzelis and Harris, p. 86).
413 TC, 424–5; Symeon Magister, 136.73–4; Skylitzes 229–30 (trans. Wortley, p. 221).
414 Liudprand of Cremona 2.15.

7 The *Sylloge Tacticorum* after the *Sylloge Tacticorum*

We have already seen that the author of the *ST* used a number of works as his sources, but that which seems to have influenced the *ST* the most was *LT*. *LT* was used as a base from which Byzantine tactics were gradually evolved and developed to correspond to contemporary needs. Therefore, the *ST* is the first manual to introduce us to what we regard as tenth-century warfare, and it is the middle point between *LT* and the *PM*. The latter will become more obvious here, since we are going to discuss the extent to which the *ST* influenced later manuals and warfare. In addition, we will look into the relationship of the *ST* with these later manuals and how tactics were further evolved and crystallized in them. The manual which seems to have been influenced the most by the *ST* is the *PM* of Nikephoros II Phokas.[1] Nikephoros II, in fact, seems to have used the *ST* in the same way that the author of the latter used *LT*. Nikephoros used the *ST* as a basis to further evolve and crystallize Byzantine armament, cavalry and infantry tactics, and the approximately thirty years which separate the *ST* from the *PM* are sufficient to explain both the similarities and the differences between the two manuals.

Developments in the armament of Byzantine troops

In terms of armament, the *PM* kept the advice of the *ST* with very few changes. With regard to the cavalry, the equipment of the Byzantine regular cavalry, the lancers, is given in more detail in the *PM*, but it is very much in accordance with the *ST*, which instructed to keep them less armoured than the *kataphraktoi* and with unarmoured horses. Likewise, judging by the equipment proposed by Nikephoros II for the horse archers in the wedge formation, the equipment of the light cavalry remained largely unchanged. The equipment of the *kataphraktoi*, however, had undergone some more important changes in the *PM*. Contrary to the *ST*, the mace became their primary weapon in the *PM*, and the use of the lance was only restricted to a small number of *kataphraktoi* in the flanks of the wedge formation. On the same grounds, the bows, which were part of their armament in the *ST*, do not appear in the *PM*, and the latter only recommended the use of *klibania* as armour, excluding the other types, such as the mail coats, that the *ST* also recorded.[2]

The armament of the infantry is more or less along the same lines. First of all, the *PM* preserves the same three infantry units as the *ST*: the heavy infantry, the medium infantry and the light infantry. The only difference is that in the *PM* infantry units are organized in *taxiarchiai* and their leader, the *taxiarchos*, is now a distinct and official rank.[3] The men are equipped in much the same way as their counterparts in the *ST*, save for the fact that Nikephoros II usually excludes lamellar and mail armour and proposes instead the cheaper types, made of felt or coarse silk and cotton.[4] Specialized infantry troops like the *menavlatoi* appear with very much the same equipment, and in fact the specific passage of the *PM* bears a strong resemblance to that of *ST*.[5]

Developments in Byzantine cavalry tactics

Apart from armament, the most interesting developments can be spotted in the evolution of tactics. It seems that Nikephoros II used the tactics of the *ST* as a template, and in his *PM* one can notice the sophistication, crystallization and improvement of both cavalry and infantry tactics.

As far as the cavalry is concerned, at first sight there seems to be very little change. Nikephoros II clearly based his instructions on the second cavalry formation found in the *ST*: both manuals record the use of irregular scouts, the *prokoursatores*, which marched ahead of the main army and were accompanied by a small *tagma* of defenders. In both cases the vanguard consisted of two units of regular cavalry in the flanks and the wedge of the *kataphraktoi* in the middle, while flank guards and out-flankers were placed even farther away in the flanks. The only difference between the two manuals so far is that the *ST* instructed that two concealed units be placed even farther away from the flank guards and out-flankers, while the *PM* omits them altogether. The second line was divided into four main units, with the general placed in the middle interval, the only difference being that the *ST* instructed another three small units to be placed in the larger intervals of this line in order to guard them. Other than that, the rest of the cavalry formation remained the same. Both manuals record a third line which was to be identical to the vanguard, if there were enough *kataphraktoi* at hand, and a rear guard of three small units followed behind.[6]

A closer look, however, is enough to spot a range of more significant changes and to indicate the optimization of cavalry tactics. These improvements are related to the use of heavy cavalry and seem to have been implemented with the aim of ensuring its success. To begin with, in both manuals, the wedge formation included *kataphraktoi*, horse archers and lancers. The *PM* is once more evidently influenced by the *ST*. Nikephoros II based his wedge in the second and third variations of the wedge of the *ST*, and so both manuals record a wedge formation of either 504 or 384 men.[7]

While the basic pattern of the wedge remained unchanged, namely consisting of twelve ranks, the men in the first rank being the fewest and every rank having additionally two men in the right and two in the left, the placement of the troops was different. In the *ST* the *kataphraktoi* covered the first, second, third, fourth, ninth, tenth, eleventh and twelfth ranks, while the other four (fifth, sixth, seventh and eighth) were filled with horse archers and lancers. It is safe to assume that as in the *PM*, the horse archers were drawn up in the middle ranks in order to be protected by the lancers.[8] Although this formation was double-faced and quite secure from encirclements, the flanks of the wedge in the middle ranks seem rather exposed. The *ST* does not clearly specify the armament of the regular cavalry, the lancers; it does state, though, that their armament should be similar to the *kataphraktoi*, but less heavy, and that their horses should be unarmoured in order to retreat and charge rapidly.[9] This fact must have made the wedge quite vulnerable, as the horses of the middle ranks could have been more easily killed by enemy troops who wished to disturb the charge of the *kataphraktoi*.

This seems to have been a really challenging issue, which the Arab cavalry probably took advantage of, since in the *PM* we can spot two counter-measures to deal with this problem. First, the wedge of the *PM* had *kataphraktoi* in all ranks. Therefore, in the middle of the wedge, there were horse archers enclosed and protected by *kataphraktoi*.[10] Although this formation was not double-faced, as the last rank was composed of both *kataphraktoi* and horse archers, its flanks were protected by the fact that all the cavalrymen in the flanks were *kataphraktoi* and thus were heavily equipped and had armoured horses. Second, the *ST* instructed the wedge to charge right after the *prokoursatores* had released their arrows against the enemy and after they had retreated through the intervals of the first line. In the *PM*, however, some of the *prokoursatores* were employed to accompany and screen the flanks of the wedge while it was charging in order to 'keep the enemy away from their flanks so that they do not divert or disrupt the *kataphraktoi* and break up the charge'. What is more, Nikephoros II provided additional instructions in case there were a great number of enemy soldiers attempting to strike against the flanks of the wedge, which unfortunately are lost today.[11]

Another optimization in the use of heavy cavalry is the abstinence from conducting pursuits. The author of the *ST* fails to notice such an important matter and seems to imply that the *kataphraktoi* were to advance towards the retreating enemy, accompanied by the other two units of regular cavalry.[12] Examples from antiquity underline the danger of this practice. In 272, the Roman emperor Aurelian lured the Palmyran *cataphracts* into pursuing the much lighter Roman cavalry, only to wheel about and attack them when they were exhausted from the pursuit. Likewise, during his western campaign (356–358), Julian the Apostate avoided pursuing the enemy with his *cataphractarii*, even though he had easily overrun them.[13] The *PM*, on the other hand, specifically instructed the general not to undertake the pursuit with the *kataphraktoi*, but instead advised to have them follow slowly as the other two accompanying units of regular cavalry conducted the pursuit.[14]

Last but not least, Nikephoros II seems to have recognized that the *kataphraktoi* were mostly unable to charge more than once. Sources from antiquity seem to support that, as Nazarius informs us that in the battle of Turin (312) the *clibanarii* were defeated after their initial charge had failed because they could not easily turn about to regroup.[15] Therefore, the *PM* provided additional advice on how to get the best out of the *kataphraktoi* and instructed the general to aim the charge specifically at the enemy commander with a steady pace and in complete silence to ensure success.[16] On the other hand, while the author of the *ST* recognized that the *kataphraktoi* could get weary from fighting, he did not give specific instructions for limiting their use against a specific target. Instead, he generally advised that enemy formation be broken with an irresistible charge but without giving any details on the procedure of the charge.

The absence of such important details from the *ST* may imply that heavy cavalry tactics were still at a preliminary stage. Indeed, two very important factors which greatly contributed to the success or failure of such specialized heavy cavalry, namely cohesion and fatigue, seem to be neglected in the *ST*. It could be that some of the experiences of the past had been forgotten due to the fact that the practice of employing specialized heavy cavalry had fallen out of use for almost four centuries. Be that as it may, the Byzantines seem to have quickly learned from their mistakes. After three decades the *PM* advanced and optimized the tactics of the *kataphraktoi* to ensure their maximum performance and to enhance their contribution in the successful operations against the Arabs.

Developments in Byzantine infantry tactics

In addition to the optimization of cavalry tactics, the *PM* records the development of infantry tactics. To begin with, Nikephoros II took measures in order to ensure the success of the specialized infantry. In the *ST* the tactics of the *menavlatoi* seem to reflect a small number of enemy heavy cavalry and a rather small threat. Their numbers were relatively small, 300 at best, and they were instructed to fight almost independently, drawing up at a distance of thirty to forty fathoms away from the infantry square.[17] In the *PM* the *menavlatoi* no longer acted alone but were part of the infantry *taxiarchia* attached to the infantry square. Each *taxiarchia* had 100 *menavlatoi*, which made them 1,200 in total, a number four times bigger than that of the *ST*.[18]

In addition, the *PM* features an optimized version of the hollow-square formation. Compared to the *ST*, Nikephoros II provided clear and practical advice on the array of the square and on counter-measures for special occasions, which shows a good understanding and crystallization of practices.[19] In this case, however, Nikephoros II did not directly advance the advice of the *ST*, but he partly copied from another source, the *Syntaxis Armatorum Quadrata*, which had already improved the hollow-square formation. Since there is confusion in the understanding of the *ST* and the *Syntaxis* and their material is falsely regarded as identical, it is essential to discuss the relationship of the *ST* and the *Syntaxis* in more detail so as to highlight the different stages and the evolution of the infantry square formation.[20]

The *Syntaxis Armatorum Quadrata* is a tactical blueprint which preserves a diagram depicting a hollow-square infantry formation with the cavalry inside. The *Syntaxis* was at some stage interpolated in the manuscript tradition of Aelian and therefore appears under his name. The diagram is accompanied by a small text which explains the formation. The text reports that there should be three infantry units in each side, twelve in total, each drawn up with an interval between each other. The intervals must have sufficient length so as to allow fifteen cavalrymen abreast to pass through. The text goes on to explain that in case of emergencies the intervals can be reduced from twelve to eight, and judging by the diagram this was achieved by merging the infantry units in the corners.[21]

The exact dating of the *Syntaxis* remains uncertain. This interpolation is usually dated before the *ST*, and since the latter is usually put c. 950, the *Syntaxis* is thought to have been compiled in first half of the tenth century, around the time of John Kourkouas.[22] Save for the exact dating, the argument that the *Syntaxis* was written before the *ST* seems convincing at first. The author of *ST* seems to have known this text, for he appears to cite it in his discussion of the hollow-square formation, stating that 'if the enemy force is larger, then it is safer, according to Polybius and Aelian, for this compound formation to have only eight intervals and the corners of the square to be attached together'.[23] In addition as we have already seen, another text interpolated in the tradition of Aelian, the *Peri Metron*, seems to have been used as a source by the author of the *ST*.[24]

When we put this thesis under closer scrutiny, however, a very important issue arises. It seems improbable for the *Syntaxis* to have influenced the *ST* since the two present different tactics. The hollow square of the *ST* seems to be in its first stage of evolution. Although the author of the *ST* tried to be flexible and adapt the shape and size of the square according to manpower and geography, he presents it in a rather theoretical and mathematical model, which is reminiscent of ancient theoretical tacticians, like Asclepiodotus and Aelian.[25] What is more, the units at the flanks of the square are not drawn up by files, as one would expect, but by ranks, which evidently makes the formation exposed from the flanks. It is hard to imagine that it would have effectively sustained an attack from the sides. It seems as if the square was not yet crystallized, but instead gradually evolved from a marching formation. On the other hand, the hollow square of the *Syntaxis* is clearly optimized. The text seems rather practical, and the units on the flanks are drawn up by files, effectively protecting the flanks of the square to the same degree as the units on the front and the back side did. Therefore, the *Syntaxis* presents an evolved and optimized model which is much more common to the one in the *PM* rather than to the *ST*. In this light, it is reasonable to argue that the *Syntaxis* was compiled sometime after the *ST* and reign of Romanos I but before the drafting of the *PM* since it was used by Nikephoros II as a source.[26]

With regard to the reference of the *ST* to Polybius and Aelian, it can be explained in another way. The reference to Polybius cannot be traced to any extant work, but it could have referred to a lost military treatise written by him.[27] The reference to Aelian is a little more complicated, and there are three possible explanations for it: the first is that the reference is fake and the author of the *ST* manipulated the tradition in order to fabricate authority for the relatively new practices he was describing.[28] The second is that the author of the *ST* did cite an interpolation of Aelian, but an older lost version of it. This is supported, as we have seen earlier, by the fact that the metrological table in the *ST* is different from the extant *Peri Metron*, which was also part of the same interpolation as the *Syntaxis*. This could therefore mean that our author drew his material from a lost version which could have included a slightly different metrological table and some other square array, perhaps an emergency or marching formation that suggested the reduction of intervals.[29] The third explanation is that the reference to Aelian was added later by the redactors of the *ST*, sometime after the *Syntaxis* had been compiled.

Even if the infantry square of the *ST* had its weaknesses, its legacy seems to have survived for at least a century and a half. We have already seen how the Byzantines could have employed such formations in the late eleventh century, as Anna Komnene speaks of a hollow square infantry formation employed by Alexios I. In addition, the influence of the *ST* could well have gone beyond Byzantium; the square fighting march formation of the Crusaders seems to have been inspired by Byzantine tactics. Since it deployed the men by ranks, the *ST* seems to be the closest Byzantine example one could find. The Crusaders could have acquired this knowledge from their previous experience as mercenaries of Byzantium or from military advice that Alexios gave to them.[30]

The last work to concern us is the *TNO*. Writing in the reign of Basil II (975–1025), Ouranos largely copied from the *PM*; however, he further revised and developed the tactics.[31] The structure of the *TNO* is very similar to the *ST*. Both treat the ideal general and more conventional warfare in their initial chapters and then move on to war by other means based on Julian Africanus and the stratagems of commanders of antiquity. Unfortunately, the manual has not yet been edited in its entirety, but so far, there is no consensus as to whether Ouranos used the *ST*. It could have been that Nikephoros Ouranos copied from the *ST* directly or that he derived his material from the non-extant *Corpus Perditum* as well as from some version of *LT*.[32]

Conclusion

The tactics of the *ST* seem to have had a quite significant impact on later manuals. The *Syntaxis* and the *PM* seem to have used the *ST* as a basis from which they optimized armament, the hollow square formation, the cavalry formation and the tactics of specialized units such as the *kataphraktoi* and the *menavlatoi*. Around the eleventh century, the *TNO* took the lead to further advance Byzantine tactics, which despite improvements, remained essentially very similar to the *ST*. It seems in fact that some of the older tactics of the *ST* were still considered practical, since they were successfully employed by the Crusaders.

Notes

1 McGeer 1995a: 184–8; Sullivan 2010a: 156–7.
2 *ST*, 39; *PM*, 3–4; McGeer 1995a: 212–17; Dawson 2002: 81–90.
3 *PM*, 1–2.1, 1.51–62,82–7,95–7; Oikonomides 1972: 273, 335–6; Kühn 1991: 273–8; McGeer 1995a: 202–11.
4 *PM*, 1–2.1; Kolias 1988: 54–8, 85–7; McGeer 1995a: 204–6; Dawson 2002: 81–90; Grotowski 2010: 166–70.
5 *ST*, 38.3; *PM*. 1.119–24; McGeer 1988: 137.
6 *ST*, 46.1–28; *PM*, 4.1–75; McGeer 1995a: 280–9.
7 *ST*, 46.26, 29; McGeer 1995a: 287–8.

8 *ST*, 46.6–8.
9 *ST*, 39.7.
10 *PM,* 3.53–65; McGeer 1995a: 286.
11 *ST,* 46.9; *PM,* 4.126–33 (trans. McGeer, p. 47); McGeer 1995a: 303–5.
12 *ST,* 46.9, 16.
13 Zosimus, 1.50; Festus, 24; Ammianus, 16.2.5–6.
14 *PM,* 4.150–3.
15 Nazarius, 22–4.
16 *PM,* 4.121–4; McGeer 1995a: 289.
17 *ST,* 47.16.
18 *PM*, 1.80–4, 94–106; McGeer 1988: 136–45; 1995a: 273–5.
19 McGeer 1995a: 262.
20 McGeer 1995a: 257–66; 1992: 223–8.
21 *ST*, 47.20; Dain 1946: 156–8; Dain and Foucault 1967: 367; McGeer 1992: 220–2.
22 McGeer 1992: 228.
23 *ST,* 47.20 (trans. Chatzelis and Harris, p. 78); McGeer 1992: 226–7.
24 See Chapter 2, p. 31.
25 *ST*, 47.24, 27; Aelian, 8–9, 15–16, 20; Rance 1994: 77.
26 Dain 1946: 159–60, proposed a similar dating theory.
27 Vári 1927: 266.
28 For a similar fabrication in religious texts see Berger 2013: 247–58. For a manipulation of Xenophon's *Anabasis* to discretely convey criticism and a moral message to Leo VI, see: Kaldellis 2015b: 22–34.
29 See Chapter 2, p. 31.
30 See Chapter 6, p. 136.
31 McGeer 1991: 132–4; Sullivan 2010a: 159–60; Trombley 1997: 271–3.
32 McGeer 1991: 132, n.27; c.f. Vieillefond 1932: lii; Dain 1937: 47–87; Foucault 1973: 281–3; Mecella 2009: 105–6.

Conclusion

The tenth century was a century of change and relative prosperity. The procedures and challenges that were introduced during the 'Dark Ages', from the Arab conquests to the ninth century, were finally crystallized and standardized. All gradual developments of Byzantine society started to accelerate. The steady revival of literacy and letters of the ninth century was further developed and facilitated in the tenth. The Byzantines, despite their defeats, had found a way to deal with the Arab threat, which was now only seasonal. After Byzantium dealt with the Bulgarians in the west, a new spirit of confidence emerged, and the initiative and offensive were undertaken in the east. The Byzantine army invaded and raided Arab lands more frequently and ever deeper. To meet the challenge effectively, the army was required to change from one that mainly focused on defensive guerrilla warfare to an army that could also field specialized troops and draw up in specialized formations, able to engage in sieges and pitched battles. These challenges, together with the so-called 'Macedonian renaissance', facilitated the production of various new military manuals which aimed at preparing the army for guerrilla warfare, sieges and pitched battles. It is in this particular context that the *ST* was produced, and one could say that its contents are very much in line with contemporary needs. This view, however, is not a foregone conclusion.

The fact that Byzantine literature often imitated classical and earlier Byzantine works sits awkwardly with modern perceptions of originality, practicality and utility. Viewed with the goggles of today's literary conventions, Byzantine literature may seem stuck in the classical tradition, with little interest in updating it and in addressing contemporary realities and needs. If we attempt to merely identify which earlier works a Byzantine author had used without first aiming to understand the purpose of *mimesis* and the authoritative role which the past played in the Middle Ages, we will end up seeing Byzantine literature as antiquarian and distortive. Byzantine military manuals, and by extension the *ST*, are by no means uninfluenced by such biases. These issues, together with the fact that the *ST* drew on a plethora of earlier works and was falsely attributed to Leo VI, created a tone of distrust and untrustworthiness for the manual, which was seen as a work copied carelessly and verbatim from two lost sources with little to no relevance for tenth-century Byzantium.

Nevertheless, an attempt to look at Byzantine literature anew with a fresh perspective and, as far as possible, prejudice-free from modern perceptions is enough to turn this view around. If we focus on inter-textual relationships between works and their models, we will realize that the Byzantines adapted and manipulated the tradition to better meet Byzantine needs and contexts. Literary conventions, secular education and the authoritative role of the past dictated that contemporary practices be veiled with a classical tunic and seen as a gradual evolution of the tradition. From an early age, the future members of the court, bureaucracy, civil and military aristocracy, as well as the throne, usually received an education which instructed them to paraphrase and adapt classical models. Against this educational and literary background, both the patrons and the authors of military treatises had much to gain from imitation: the commissioners and patrons of military manuals, who were usually emperors, would have been seen as protectors and continuators of the classical wisdom, while the authors would have gained authority for their content, a more refined style and a chance to demonstrate their erudition.

In this light, the author of the *ST* was just a regular example of an established tradition. For his sources he used various works from different genres: classical and earlier Byzantine military manuals, Byzantine *anthologia*, mirrors of princes and legal texts, as well as treatises which included recipes for the production of poisons, antidotes, flammable and non-flammable mixtures. The author of the *ST* does not seem to have had direct access to all of these sources. Some of them, mostly those that appear in the second half of the work (chapters 56–102), were copied from a non-extant source, the *Corpus Perditum*. Although sometimes it is necessary to assume that some lost work existed, there is a risk of overdoing it. Assuming that a lost source existed every time a passage was adapted can silence important aspects of style and authorship within extant narratives. Instead, it seems more probable that the author of the *ST* copied the majority of his sources from individual extant works, which he adapted and paraphrased. Therefore, the author of the *ST* was by no means an extraordinary example of slavish copying, and his narrative is at least as trustworthy as that of other tenth-century manuals such as *LT* or the *PM*.

With this in mind, one can safely turn to the internal evidence of the *ST* in order to determine its dating. The testimony of the *ST* speaks of any army ready to fulfil the traditional defensive guerrilla role, as well as an offensive one: an army which could employ specialized troops and sophisticated formations. In this light, the reign of Romanos I is by far the most likely period of composition not only because it is when the Byzantines started to take the initiative but also because the army of the *ST*, technologically, administratively and operationally, fits into this period. The tactics, ranks and technology paint the picture of an army which had gradually evolved from the reign of Leo VI but not so much as to be dated to the reign of Constantine VII. Last but not least, it is not only that these developments occurred and fit the context of the reign of Romanos I but also that they are compatible with Arab developments of the first half of the tenth century and that they can be seen and explained as a response to these enemy developments.

The authorship of Romanos I is also supported by the fact that the *ST* does not refer to members of the Macedonian dynasty, even though its sources seem to have included such references. It was a characteristic of the Macedonian dynasty to practise imperial propaganda through literature, and both Leo VI and Constantine VII refer to their predecessors in their works. For Romanos I, however, it would have been a necessity to avoid any memoranda that linked to the Macedonian dynasty and to Constantine VII, who was the legitimate heir to the throne. The false attribution of the *ST* could well have been a *damnatio memoriae* imposed by Constantine VII and Basil Lekapenos. This is not, after all, something unpreceded. Constantine VII used a similar action against Alexander after the latter had died. In addition, when Romanos I was still on the throne, Constantine VII spoke highly of him, but after he was dethroned, Constantine VII stained his reputation at any given chance. Other men in the circle of Constantine VII, either the Gongylioi brothers or Basil Lekapenos, were also hostile to the name of Romanos I, as well as to his closest associates, imposing a *damnatio memoriae* on his *domestikos* of the *scholai*, John Kourkouas, and on his *parakoimomenos*, Theophanes.

Constantine VII and Basil Lekapenos seem not only to have changed the attribution of the *ST* but also to have further interfered with the text. The method of the revision of the *ST* has many similarities with the way Constantine VII and Basil Lekapenos revised other extant treatises. Given that there is no alternative version of the *ST* available, only a few conjectures can be made as to the level of their editing. Judging from their usual method, they seem to have re-ordered the material in a certain way, adding new chapters and new chapter headings. They might have also added or gathered together in one section all the stratagems in the last part of the manual, as well as written the *scholion* that appears after chapter 46. As it was usually the case with the works they revised, the inconsistencies in the introduction and in the cross-references were not corrected to reflect the changes, and one chapter from the stratagems is also missing.

The inconsistent cross-references, as well as the uncertain dating and attribution, greatly contributed to the image of the *ST* as an untrustworthy source. But since the original compiler of the *ST* was not responsible for them and the testimony of the *ST* is as credible as any other tenth-century Byzantine manual, it is time to re-assess and re-evaluate its contents. The contribution of the *ST* to our understanding of the tenth century is significant. It is a manual which provides insights into how the Byzantines interacted with the past and how they updated classical sources to fit them into their own contemporary context. An inter-textual analysis contributes to our understanding of how the Byzantines perceived war and their enemies and how Christianity made some classical passages obsolete. What is more, the *ST* is a source which perfectly reflects the ideology and values of military aristocracy as these had begun to take shape in the tenth century.

But apart from the ideological, cultural and literary sphere, the greatest contribution of the *ST* is its military innovations. The army of the *ST* is an army that had gradually evolved from *LT*. It is an army highly specialized and divided into light, medium and heavy infantry and cavalry, with updated equipment and technology. The troops were drawn up in new formations which were a more sophisticated version of the practices found in *LT*. All the aspects which introduce us to tenth-century strategy, tactics and equipment as we know them were first recorded in the *ST*.

The fact that an author has updated the tradition and has included original material in his manual, however, does not necessarily mean that this material was practical and applicable on the battlefield. One may see military manuals primarily as literary projects of more interest to scholars and politicians rather than generals. In this light the main purpose for their production could have been to experiment linguistically and philologically, to preserve the classical tradition and to appear as its protector and continuator, as well as to seek political advancement, promotion and legitimacy.

Although these elements are present in military manuals, they do not seem to have been central. Practicality, in fact, seems compatible with some of these elements, but at the end of the day, it would be futile to try to understand the purpose of military manuals without first looking into the role that tradition and history played in shaping the thinking and worldview of the Byzantine aristocracy. There is, without doubt, more than one motive for the compilation of any work. An author could have certainly produced a work to promote himself among his peers and to appear as the continuator of classical wisdom, but what better way for someone to advertise himself as the best candidate for power by producing the most up-to-date work on military matters. The correct way to do so, however, would have been within the traditional standards of Byzantine literature. The past played a very authoritative role in Byzantium – sometimes, in fact, it enjoyed more credibility than current observation or practice. Every innovation had to be connected with the older tradition so as to be seen as a gradual evolution from it and not as a rigid break. Therefore, the correct way to proceed would have been to paraphrase classical and older sources whenever something had already been covered by them and, when necessary, to update them with some original material. What is more, classical or older material does not necessarily mean outdated material. Most of the time, the advice of Onasander or *MS* was still relevant in the tenth century, and even more so with a little updating. The anecdotes of Polyaenus could have been of use to anyone in theory; not only did they provide exemplars from the classical past, they also supplemented and connected material from the first half of the treatise with the deeds of prominent figures from antiquity.

While it is easy to underestimate the practical use of manuals, it is also very easy to overestimate it. A parallel passage between a military manual and a historical narrative is not enough to confirm that the advice of manuals was followed on the battlefield. A great deal of our historical narratives include promotional sources which could well have presented some generals in an ideal light, while battle narratives may have been fictional. It is possible that they preserved a core of truth, but they could also have been deliberately adapted to include practices which were similar to the manuals. It is therefore necessary not to take the testimony of a source at face value before its sources and motives have been clarified. The only way to confirm to what extent the advice of manuals was followed is to compare Byzantine historical narratives with other types of sources, such as administrative or legal records, as well as with Arab, Western or Armenian material, when possible.

Generally speaking, most of the material of the *ST* seems to be in line with tenth-century context and practice. In this line of reasoning most of its contents seems practical. It is indeed a different issue if a manual had relevant and rational advice to offer, and a different issue if people actually employed its practices on the battlefield. To what extent prominent generals actually followed the advice of manuals is something which has to do with how fictional promotional narratives are. When we do have parallel accounts, it seems that most of the information of the *ST* was considered common practice and that promotional sources did not fabricate events out of thin air. They most certainly enhanced the image of their subjects with allusions, anecdotes and praise, but the basic strategies followed do not seem to have been invented.

After all, the advice of the *ST* was not discarded as theoretical, and this is evident by the afterlife of the manual. Very experienced generals who wrote military manuals in the tenth and eleventh centuries, like Nikephoros II Phokas and Nikephoros Ouranos, used the information of the *ST* as a basis. It is true that they further advanced and sophisticated Byzantine tactics, but the revision was not drastic. At their core, Byzantine tactics in the eleventh century, as reflected in the *TNO*, remained largely the same as those in the *ST*. The advice of the *ST* seems not only to have been appreciated by the Byzantines but by Westerners as well. The Crusaders fought their way through to Ascalon by employing similar tactics as the *ST* describes.

Bibliography

Editions of the *Sylloge Tacticorum*

Köchly, H. (ed.). 1854. *Index lectionum in literarum universitate Turicensi habendarum*, 2 vols., Zurich.

Migne, J.P. (ed.). 1863. *Patrologia Graeca*, vol. 107, Paris.

Melber, J. (ed.). 1887. *Polyaeni Strategematon libri octo*, Leipzig.

Vári, R. (ed.). 1917–1922. *Leonis imperatoris Tactica*, 2 vols., Bucharest.

Vieillefond, J.R. (ed.). 1932. *Jules Africain, fragments des Cestes provenant de la collection des tacticiens grecs*, Paris.

Dain, A. (ed.). 1938. *Sylloge tacticorum, quae olim "Inedita Leonis tactica" dicebatur*, Paris.

Primary sources

Aelian. 2012. *The Tactics of Aelian, or On the Military Arrangements of the Greeks*, ed. and trans. C. Matthew, Barnsley.

Aeneas the Tactician. 1967. *Énée le tacticien. Poliorcétique*, ed. A.M. Bon and A. Dain, Paris.

Agathias Scholasticus. 1967. *Agathiae Myrinaei Historiarum libri quinque*, ed. R. Keydell, Berlin; Eng. trans. J.D. Frendo, *Agathias the Histories*, Berlin, 1975.

Albert of Aachen. 2007. *Albert of Aachen Historia ierosolimitana, History of the Journey to Jerusalem*, ed. and trans. S.B. Edgington, Oxford.

Ammianus Marcellinus. 1935–1939. *Ammiani Marcellini Rerum Gestarum*, ed. and trans. J.C. Rolfe, 3 vols., London and Cambridge, MA.

Apparatus Bellicus. 1746. In: I. Lamis (ed.), *Ioannis Meursi Operum, Volumen Septimum*, Florence: 905–84.

Arrian. 1976–1983. *Arrian, Anabasis of Alexander*, ed. and trans. P.A. Brunt, 2 vols., Cambridge, MA.

Attaleiates Michael. 2012. *Michael Attaleiates: The History*, ed. and trans. A. Kaldellis and D. Krallis, Cambridge, MA and London.

Basil I. 2009. *Βασίλειος Α' Μακεδών: Δύο παραινετικά κείμενα προς τον Λέοντα ΣΤ'*, ed. and trans. K. Païdas, Athens.

Basil the Younger. 2014. *The life of St. Basil the Younger: Critical Edition and Annotated Translation of the Moscow Version*, ed. and trans. D.F. Sullivan, et al., Washington, DC.

Bryennios Nikephoros. 1975. *Nicéphore Bryennios Histoire*, ed. and trans. P. Gautier, Brussels.

Chrysostom, John. 1912. *In Omnes Sanctos*. In: C.I. Dyobouniotes (ed.), Ὁ ὑπ᾽ἀριθμ. 108 κῶδιξ τῆς Ἱ. Συνόδου τῆς Ἐκκλησιας τῆς Ἑλλάδος᾽. *Ἐκκλησιαστικός Φάρος* 9: 303–5.

Chrysostom, John. 1957. *In Sanctum Pascha*. In: F. Floëri and P. Nautin (eds.), *Homélies pascales*, vol. 3, Paris.

Constantine VII. 1829. *De Cerimoniis*. In: J.J. Reiske (ed.), *Constantini Porphyrogeniti imperatoris de cerimoniis aulae Byzantine libri duo*, Bonn; Eng. trans. A. Moffatt and M. Tall, *Constantine Porphyrogennetos: The Book of Ceremonies*, Canberra, 2012.

Constantine VII. 1838. 'Theophanes Continuatus'. In: I. Bekker (ed.), *Theophanes Continuatus, Ioannes Cameniata, Symeon Magister, Georgius Monachus*, Bonn: 3–481.

Constantine VII. 1952. *De Thematibus*. In: A. Pertusi (ed.), *Costantino Porfirogenito, De thematibus*, Vatican City.

Constantine VII. 1967a. *De Administrando Imperio*. In: G. Moravcsik (ed.) and R.J.H. Jenkins (trans.), *Constantine Porphyrogenitus, De administrando imperio*, Washington, DC.

Constantine VII. 1967b. *Military Oration*. In: H.G. Ahrweiler (ed.), 'Un discours inédit de Constantin VII Porphyrogénète'. *TM* 2: 393–404; Eng. trans. E. McGeer, 'Two military orations of Constantine VII'. In: J.W. Nesbitt (ed.), *Byzantine Authors: Literary Activities and Preoccupations, Texts and Translations Dedicated to the Memory of Nicolas Oikonomides*, Leiden and Boston, 2003: 117–20.

Constantine VII. 1990. *Three Treatises*. In: J.F. Haldon (ed. and trans.), *Constantine Porphyrogenitus: Three Treatises on Imperial Military Expeditions*, Vienna.

Constantine VII. 2011. *Vita Basilii*. In: I. Ševčenko (ed. and trans.), *Chronographiae quae Theophanis Continuati nomine fertur liber quo Vita Basilii imperatoris amplectitur*, Berlin.

Constantine VII. 2015. 'Theophanes Continuatus'. In: J.M. Featherstone and J.S. Codoñer (ed. and trans.), *Chronographiae quae Theophanis Continuati nomine fertur libri i–iv*, Boston and Berlin.

Corripus, Flavius. 1970. *Iohannis*. In: I. Diggle and F.R.D. Goodyear (eds.), *Flavii Cresconii Corippi, Iohannidos Libri VIII*, Cambridge; Eng. trans. W. Shea, *The Iohannis or De Bellis Libycis of Flavius Cresconius Corippus*, New York, 1998.

Corpus Parisinum. 2007. In: D.M. Searby (ed. and trans.), *The Corpus Parisinum, A Critical Edition of the Greek Text with Commentary and English Translation: A Medieval Anthology of Greek Texts from the Pre-Socratics to the Church Fathers 600 B.C.–700 A.D.*, 2 vols., New York.

Damascenus, John. 1914. *Vita Barlaam et Joasaph*. In: H. Mattingly and G.R. Woodward (eds. and trans.), *St. John Damascene, Barlaam and Joasaph*, Cambridge, MA.

Daphnopates, Theodore. 1978. *Théodore Daphnopatès correspondance*, ed. and trans. J. Darrouzès and L.G. Westerink, Paris.

De Obsidione Toleranda. 1947. In: H. van den Berg (ed.), *Anonymus De obsidione toleranda: edition critica*, Leiden; Reprint, Eng. Trans. and annot. D.F. Sullivan, 'A Byzantine instructional manual on siege defense: The *De obsidione Toleranda*. Introduction, English translation and annotations'. In: J.W. Nesbitt (ed.), *Byzantine Authors: Literary Activities and Preoccupations, Texts and Translations Dedicated to the Memory of Nicolas Oikonomides*, Leiden and Boston, 2003: 139–266.

De Re Militari. 1985. In: G.T. Dennis (ed. and trans.), *Three Byzantine Military Treatises*, Washington, DC: 246–327.

Ekloga. 1983. In: L. Burgmann (ed. and trans.), *Ecloga: Das Gesetzbuch Leonis III. und Konstantinos V*, Frankfurt.

Euthymius Protasecretis. 1981. *Encomium in sanctum Theodorum Stratelatem*. In: F. Halkin, 'L'éloge de saint Théodore de Stratélate'. *Analecta Bollandiana* 99: 223–37.

Festus. 2009. *Il Breviarium di Rufio Festo*, ed. and trans. M.L. Fele, Hildesheim.

Genesios, Joseph. 1978. *Iosephi Genesii regum libri quattuor*, ed. A. Lesmüller-Werner and J. Thurn, Berlin; Eng. trans. A. Kaldellis, *Genesios on the Reigns of the Emperors*, Canberra, 1998.

Geoponika. 2010. In: E. Lelli (ed. and trans.), *L'agricoltura antica, I Geoponica di Cassiano Basso*, Soveria Mannelli; Eng. trans. A. Dalby, *Geoponika, Farm Work*, Totnes, 2011.

Gregorius. 2007. *Commentarius in Ecclesiasten*. In: G.H. Ettlinger and J. Noret (eds.), *Pseudo-Gregorii Agrigentini seu Pseudo-Gregorii Nysseni commentaries in Ecclesiasten*, Turnhout.

Grierson, P. 1973. *Catalogue of the Byzantine Coins in the Dumbarton Oaks Collection and in the Whittemore Collection*, vol. 3.2, Washington, DC.

Heron. 1903. *Geometrica*, In: J.L. Heiberg (ed.), *Heronis Alexandrini opera quae supersunt omnia*, vol. 4, Leipzig: 172–448.

Heron of Byzantium. 2000. *Parangelmata Poliorketika*. In: D.F. Sullivan (ed. and trans.), *Siegecraft, Two Tenth-Century Instructional Manuals by "Heron of Byzantium"*, Washington, DC: 26–113.

Hypothesis. 1887. In: J. Melber (ed.), *Polyaeni Strategematon libri octo*, Leipzig: 427–500; Eng. trans. P. Krentz and E.L. Wheeler, *Polyaenus Stratagems of War*, 2 vols., Chicago, 1994: ii. 852–1003.

Isocrates. 1928a. *To Nicocles*. In: G. Norlin (ed. and trans.), *Isocrates*, 3 vols., Cambridge, MA: i.37–72.

Isocrates. 1928b. *To Demonicus*. In: G. Norlin (ed. and trans.), *Isocrates*, 3 vols., Cambridge, MA: i.1–36.

Julian the Apostate. 1913–1923. *Julian*, ed. and trans. W.C.F. Wright, 2 vols., London and Cambridge, MA.

Julian of Ascalon. 1996. *Le traité d'urbanisme de Julien d'Ascalon (VIe siècle)*, ed. and trans. C. Saliou, Parris.

Kaminiates, John. 1973. *Ioannis Caminiatae de expugnatione Thessalonicae*, ed. G. Böhlig, Berlin; Eng. trans. D. Frendo and A. Fotiou, *John Kaminiates, The Capture of Thessaloniki*, Perth, 2000.

Kekaumenos, Katakalon. 2013. *Kekaumenos, Advice and Anecdotes, Edition, English Translation and Commentary*, ed. and trans. C. Roueché, online edition. www.ancientwisdoms. ac.uk/folioscope/greekLit%3Atlg3017.Syno298.sawsGrc01

Kletorologion. 1972. In: N. Oikonomides (ed. and trans.), *Les listes de préséance byzantines des IXe et Xe siècles*, Paris: 65–235.

Komnene, Anna. 2001. *Annae Comnenae Alexias*, ed. A. Kambylis and D.R. Reinsch, Berlin and New York; Eng. trans. E.R.A. Sewter and P. Frankopan, *Anna Komnene Alexiad*, London, 2009.

Leo VI. 1944. *Novels*. In: A. Dain and P. Noialles (eds.), *Les novelles de Léon VI le Sage*, Paris.

Leo VI. 2014. *Taktika*. In: G.T. Dennis (ed. and trans.), *The Taktika of Leo VI*, revised edition, Washington, DC.

Leo the Deacon. 1828. *Leonis Diaconi Caloensis historiae libri decem*, ed. C.B. Hase, Bonn; Eng. trans. A.M. Talbot and D.F. Sullivan, *The History of Leo the Deacon, Byzantine Military Expansion in the Tenth Century*, Washington, DC, 2005.

Leo the Grammarian. 1842. *Leonis Grammatici Chronographia*, ed. I. Bekker, Bonn.

Libanius. 1903–1922. *Libanii opera*, ed. R. Foerster, 11 vols., Leipzig.

Liudprand of Cremona. 1998. *Antapodosis*. In: P. Chiesa (ed.), *Liudprandi Cremonensis Antapodosis, Homelia Paschalis, Historia Ottonis, Relatio de Legatione Constantinopolitana*, Turnhout: 3–150; Eng. trans. In: P. Squatriti, *The Complete Works of Liudprand of Cremona*, Washington, DC, 2007: 41–102.

Mas'udi. 2007. *Al Mas'udi, From the Meadows of Gold*, trans. P. Lunde and C. Stone, London.

Maurice. 1981. *Strategikon*. In: G. T. Dennis (ed.) and E. Gamillscheg (trans.), *Das Strategikon des Maurikios*, Vienna; Eng. trans. G.T. Dennis, *Maurice's Strategikon: Handbook of Byzantine Military Strategy*, Philadelphia, 1984.

McGeer, E. (trans.). 2000. *The Land Legislation of the Macedonian Emperors*, Toronto.

Mémorandum. 1940. In: A. Dain (ed. and trans.), 'Mémorandum inédit sur la défense des places'. *Revue des Études Grecques* 53: 123–36.

Migne, J.P. (ed.). 1863. *Patrologia Graeca*, vol. 35, Paris.

Miracula Demetrii. 1979. In: P. Lemerle (ed.), *Les plus anciens recueils des miracles de saint Démétrius et la pénétration des Slaves dans les Balkans*, Paris.

Miskawaihi. 1921. *The Eclipse of the Abbasid Caliphate*, eds. and trans. H.F. Amedroz and D.S. Margoliouth, vol. 4, Oxford.

Nazarius. 1874. *XII Panegyrici Latini*, ed. E. Baehrens, Leipzig; Eng. trans. C.E.V. Nixon and B.S. Rogers, *In Praise of Later Roman Emperors: The Panegyrici Latini*, Los Angeles and Oxford, 1994.

Nicomachus the Pythagorean. 1895. *Harmonicum enchiridion*. In: K. Jan (ed.), *Musici scriptores Graeci*, Leipzig: 236–65; Eng. trans. F.R. Levin, *The Manual of Harmonics of Nicomachus the Pythagorean*, Grand Rapids, MI, 1994.

Nikephoros II. 1986. *De Velitatione*. In: G. Dagron and H. Mihăescu (eds. and trans.), *Le traité sur la guérilla (De velitatione) de l'empereur Nicéphore Phocas (963–969)*, Paris; Eng. trans. 'Skirmishing'. In: G.T. Dennis (ed.), *Three Byzantine Military Treatises*, Washington, DC, 1985: 145–239.

Nikephoros II. 1995. *Praecepta Militaria*. In: E. McGeer (ed. and trans.), *Sowing the Dragon's Teeth: Byzantine Warfare in the Tenth Century*, Washington, DC: 12–59.

Nikephoros Ouranos. 1995. *Taktika*. In: E. McGeer (ed. and trans.), *Sowing the Dragon's Teeth: Byzantine Warfare in the Tenth Century*, Washington, DC: 88–163.

Onasander. 1928. In: W.A. Oldfather and C.H. Oldfather (ed. and trans.), *Aeneas Tacticus, Asclepiodotus, Onasander*, Cambridge, MA and London: 368–527.

Peri Metron. 1946. In: A. Dain (ed.), *Histoire du texte d'Élien le tactician des origines à la fin du moyen age*, Parris: 158–9.

Photios. 1822. *Lexicon*, In: R. Porson (ed.), *Φωτίου τοῦ πατριάρχου λέξεων συναγωγή*, 2 vols., Cambridge.

Poinalios Nomos. In: *JGR*: ii. 80–9.

Polyaenus. 1887. *Polyaeni strategematon libri viii*, ed. J. Melber and E. Woelfflin, Leipzig: 2–425; Eng. trans. In: P. Krentz and E.L. Wheeler (eds.), *Polyaenus Stratagems of War*, 2 vols., Chicago: i. 1–1003.

Polybius. 1922–1927. *The Histories*, ed. and trans. W.R. Paton, 6 vols., Cambridge, MA and London.

Procheiros Nomos. In: *JGR*: ii. 114–228.

Procopius. 1914–1928. *Procopius of Caesarea, History of the Wars*, ed. and trans. H.B. Dewing, 5 vols., Cambridge, MA and London.

Psellos, Michael. 1990. *Historia*. In: W.J. Aerts (ed. and trans.), *Michaeli Pselli Historia Syntomos*, Berlin.

Psellos, Michael. 2014. *Chronographia*. In: D.R. Reinsch (ed.), *Michaelis Pselli Chronographia*, Berlin and Boston; Eng. trans. E.R.A. Sewter, *Fourteen Byzantine Rulers: The "Chronographia" of Michael Psellos*, Harmondsworth, 1979.

Ralph of Caen. 2005. *The Gesta Tancredi of Ralph of Caen: A History of the Normans of the First Crusade*, trans. B.S. Bachrach and D.S. Bachrach, Aldershot and Burlington, VT.

Raymond of Aguilers. 1968. *Raymond d'Aguilers, Historia Francorum qui ceperunt Iherusalem*, trans. H. Hill and L.L. Hill, Philadelphia.

Skylitzes, John. 1973. *Ioannis Scylitzae Synopsis Historiarum*, ed. I. Thurn, Berlin and New York; Eng. trans. J. Wortley, *John Skylitzes, A Synopsis of Byzantine History, 811–1057*, Cambridge and New York, 2010.

Stephen of Taron. 1909. *Des Stephanos von Taron armenische Geschichte*, trans. H. Gelzer and A. Burckhardt, Leipzig.

Stobaeus, John. 1884–1912. *Ioannes Stobaei Anthologium*, ed. O. Hense and C. Wachmuth, 5 vols., Berlin.

Strabo. 1873–1877. *Strabonis Geographica*, ed. A. Meineke, 3 vols., Leipzig.

Stratagemata. 1949. In: J.A. de Foucault (ed.), *Stratagemata*, Paris: 17–66.

Suda. 1928–1935. *Suidae Lexicon*, ed. A. Meineke, 4 vols., Leipzig.

Sylloge Tacticorum. 1938. In: A. Dain (ed.), *Sylloge Tacticorum quae olim "Inedita Leonis tactica" dicebatur*, Paris; Eng. trans. G. Chatzelis and J. Harris, *A Tenth-Century Byzantine Military Manual: The Sylloge Tacticorum*, Abingdon and New York.

Symeon Magister. 2006. *Symeonis Magistri et Logothetae Chronicon*, ed. S. Wahlgren, Berlin.

Synagoge. 2008. In: I.C. Cunningham (ed.), *Sammlung griechischer und lateinischer Grammatiker*, Berlin and New York.

Synaxarium Mensis Septembris. 1902. In: H. Delehaye (ed.), *Acta Sanctorum*, vol. 62, Brussels: 1–94.

Syntaxis Armatorum Quadrata. 1992. In: E. McGeer (ed. and trans.), 'The Syntaxis Armatorum Quadrata: A Tenth-Century Tactical Blueprint'. *REB* 50: 219–29.

Syrianos Magister. 1985. *Peri Strategias*. In: G.T. Dennis (ed. and trans.), *Three Byzantine Military Treatises*, Washington DC 1985: 11–135.

Tabarī. 1989–2007. *The History of al-Ṭabarī*, trans. F. Rosenthal, 40 vols., New York.

Theodosius the Deacon. 1979. *Theodosii diaconi de Creta capta*, ed. H. Criscuolo, Leipzig.

Theon, Aelius. 1997. *Progymnasmata*. In: M. Patillon (ed. and trans.), *Aelius Théon Progymnasmata*, Paris; Eng. trans. In: G.A. Kennedy, *Progymnasmata: Greek Textbooks of Prose Composition and Rhetoric*, Atlanta, GA: 1–72.

Theophanes Confessor. 1883. *Theophanis chronographia*, ed. C. de Boor, Leipzig; Eng. trans. C. Mango and R. Scott, *The Chronicle of Theophanes Confessor*, Oxford, 1997.

Theophylact Simocatta. 1887. *Theophylacti Simocattae Historiae*, ed. C. de Boor, Leipzig; Eng. trans. L.M. Whitby and M. Whitby, *The History of Theophylact Simocatta*, Oxford, 1986.

Urbicius. 2005. In: R. Burgess (ed.), 'Urbicius' *Epitideuma*: An Edition, Translation and Commentary'. Eng trans. R. Burgess and G. Greatrex, *BZ* 98: 35–74.

William of Apulia. 1961. *La Geste de Robert Guiscard*, ed. and trans. M. Mathieu, Palermo; Eng. trans. G.A. Loud, *The Deeds of Robert Guiscard*, Leeds, 2002.

Yahya of Antioch. 1925. In: I. Kratchkovsky and A. Vasiliev (ed. and trans.), 'Histoire de Yahya ibn-Said d' Antioche continuateur de Said ibn-Bitriq'. *Patrologia Orientalis* 18: 700–833; I. Kratchkovsky (ed.), trans. F. Micheau and G. Tropeau, *Patrologia Orientalis* 47 (1997): 385–539.

Zepos, I. and Zepos, P. (eds.). 1931. *Jus Graecoromanum*, 8 vols., Athens.

Zosimus. 1971–1989. *Zosime Histoire Nouvelle*, ed. and trans. I.F. Paschoud, 4 vols., Paris.

Secondary sources

Ahrweiler, H.G. 1960. 'Recherches sur l'administration de l'empire byzantine aux IX-Xième siècles'. *BCH* 84: 1–111.

Ahrweiler, H.G. 1966. *Byzance et la mer: La marine de guerre, la politique et les institutions maritimes de Byzance aux Viie–XVe siècles*, Paris.

Ahrweiler, H.G. 1967. 'Un discours inédit de Constantin VII Porphyrogénète'. *TM* 2: 394–404.

Ahrweiler, H.G. 1971. 'La frontière et les frontières de Byzance en Orient'. In: M. Berza and E. Stănescu (eds.), *Actes du XIVe Congrès international des études byzantines*, vol. 1, Bucharest: 209–30.

Ahrweiler, H.G. 1981. 'Sur la date du De Thematibus de Constantin VII Porphyrogénète'. *TM* 8: 1–5.

Alexopoulos, T. 2008–2012. 'Using ancient military handbooks to fight medieval battles: Two stratagems used by Alexios I Comnenos against the Normans and Pechenegs'. *Εωα και Εσπερία* 8: 47–71.

Anagnostakis, I. 1999. 'Ούκ εΐσιν εμά τα γράμματα. Ιστορία και ιστορίες στον Πορφυρογέννητο'. *Βυζαντινά Σύμμεικτα* 13: 97–140.

Anagnostakis, I. 2008. 'Η Σολομώντεια αμφιθυμία των πρώτων Μακεδόνων αυτοκρατόρων και οι αποκαλυπτικές καταβολές της'. In: A. Lambropoulou and K. Tsiknakis (eds.), *Η εβραϊκή παρουσία στον ελλαδικό χώρο (4ος–19ος αι.)*, Athens: 39–60.

Anastasiadis, M.P. 1994. 'On handling the menavlion'. *BMGS* 18: 1–10.

Andriollo, L. 2012. 'Les Kourkouas'. *Studies in Byzantine Sigillography* 11: 57–88.

Andriollo, L. 2014. 'Aristocracy and literary production in the 10th century'. In: A. Pizzone (ed.), *The Author in Middle Byzantine Literature: Modes, Functions and Identities*, Berlin: 119–38.

Andriollo, L. 2017. *Constantinople et les provinces d'Asie Mineure IXe–XIe siècles: Administration impériale, sociétés locales et role de l'aristocratie*, Leuven, Paris, and Bristol, CT.

Andrist, P. 2007. *Les manuscrits grecs conservés à la Bibliothèque de la Bourgeoisie de Berne – Burgerbibliothek Bern: Catalogue et histoire de la collection*, Zurich.

Angelidi, C. 1980. *Ο βίος του οσίου Βασιλείου του Νέου*, PhD thesis, University of Ioannina, Ioannina.

Angelidi, C. 2013. 'Basil Lecapène. "Deux ou trois choses que je sais de lui"'. In: C. Gastgeber et al. (eds.), *Pour l'amour de Byzance: Hommage à Paolo Odorico*, Frankfurt: 11–26.

Apostolopoulou, S. 1982. 'Η άλωση της Μοψουεστίας (965) και της Ταρσού (965) από βυζαντινές και αραβικές πηγές'. *Graeco-Arabica* 1: 157–67.

Asa Eger, A. 2011. *The Spaces between the Teeth: A Gazetteer of Towns on the Islamic-Byzantine Frontier*, Istanbul.

Asa Eger, A. 2014. *The Islamic-Byzantine Frontier: Interaction and Exchange among Muslim and Christian Communities*, London and New York.

Ashburner, W. 1926. 'The Byzantine Mutiny Act'. *The Journal of Hellenic Studies* 46: 80–109.

Ayalon, D. 1994. 'The military reforms of caliph al-Muʿtaṣim, their background and consequences'. In: *Islam and the Abode of War*, Aldershot and Burlington, VT: 1–39.

Bandini, A.M. 1770. *Catalogus codicum manuscriptorum bibliothecae Mediceae Laurentianae*, vol. 3, Florence.

Belke, K., Hellenkemper, H., Hild, F., Koder, J., Küzler, A., Merisch, N., Restle, M., Soustal, P., Todt, K.-P. and Vest, B.A. 1976–2014. *Tabula Imperii Byzantini*, 15 vols., Vienna.

Bennett, M. 2001. 'The crusaders' fighting march revisited'. *War in History* 8: 1–18.

Berger, A. 2013. 'Believe it or not: Authority in religious texts'. In: P. Armstrong (ed.), *Authority in Byzantium*, Farnham and Burlington, VT: 247–58.

Beihammer, A. 2012. 'Strategies of diplomacy and ambassadors in Byzantine-Muslim relations on the tenth and eleventh centuries'. In: A. Becker and N. Drocourt (eds.), *Ambassadeurs et ambassades au cœur des relations diplomatiques: Rome – Occident medieval – Byzance (VIIIe s. Avant J.-C.–XIIe s. Après J.-C.)*, Metz: 371–400.

Bianconi, D. 2008. 'La controversia palamitica: Figure, libri, testi e mani'. *Segno e Testo* 6: 337–76.

Bianconi, D. 2011. 'Un altro Plutarco di Planude'. *Segno e Testo* 9: 113–26.

Bianconi, D. 2012. '"Duplici scribendi forma". Commentare Bernard de Montfaucon'. *Medioevo e Rinascimento* 23: 299–317.

Bikhazi, R. 1981. *The Hamdanid Dynasty of Mesopotamia and North Syria*, PhD thesis, University of Michigan, Ann Arbor, MI.

Birkenmeier, J.W. 2002. *The Development of the Komnenian Army*, Leiden and Boston.

Bivar, A.D.H. 1972. 'Cavalry equipment and tactics on the Euphrates frontier'. *DOP* 26: 271–91.

Bonner, M. 1996. *Aristocratic Violence and Holy War: Studies in the Jihad and the Arab-Byzantine Frontier*, New Haven, CT.

Bosworth, C.E. 1965–1966. 'Military organisation under the Buyids of Persia and Iraq'. *Oriens* 18/19: 143–67.

Bosworth, C.E. 1992. 'The city of Tarsus and the Arab-Byzantine frontiers in early and middle Abbāsid times'. *Oriens* 33: 268–86.

Brokkaar, W.G. 1972. 'Basil Lecapenus: Byzantium in the tenth century'. In: W.F. Bakker et al. (eds.), *Studia Byzantina et Neohellenica Neerlandica*, Leiden: 199–234.

Brown, P. 2011. 'The Gesta Roberti Wiscardi: A "Byzantine" history?'. *Journal of Medieval History* 37: 162–79.

Browning, R. 1978. 'Literacy in the Byzantine world'. *BMGS* 4: 39–54.

Buckley, P. 2006. 'War and peace in the Alexiad'. In: J. Burke (ed.), *Byzantine Narrative*, Melbourne: 92–109.

Buckley, P. 2014. *The Alexiad of Anna Komnene: Artistic Strategy in the Making of a Myth*, Cambridge.

Bury, J.B. 1906. 'The treatise De administrando imperio'. *BZ* 15: 517–77.

Bury, J.B. 1907. 'The ceremonial books of Constantine Porphyrogennetos'. *English Historical Review* 86/87: 209–27, 417–39.

Bury, J.B. 1909. 'Mutasim's march through Cappadocia in A.D. 838'. *Journal of Hellenic Studies* 29: 120–9.

Canard, M. 1951. *Hamdanides: Histoire de la dynastie des Hamdanides de Jazîra et de Syrie*, Algiers.

Cappel, A.J. 1994. 'The Byzantine response to the Arabs (10th–11th centuries)'. *BF* 20: 113–32.

Chatzelis, G. and Harris, J. 2017. 'Introduction and notes'. In: *A Tenth-Century Byzantine Military Manual: The Sylloge Tacticorum*, Abingdon and New York.

Chernoglazov, D. 2013. 'Beobachtungen zu den Briefen des Theodoros Daphnopates. Neue Tendenzen in der byzantinischen Literatur des zehnten Jahrhunderts'. *BZ* 106: 623–44.

Chevedden, P.E. 2000. 'The invention of the counterweight trebuchet: A study in cultural diffusion'. *DOP* 54: 71–116.

Cheynet, J.C. 1986a. 'Les Phocas'. In: G. Dagron and H. Mihăescu (eds.), *Le traité sur la guérrilla (De velitatione) de l'empereur Nicéphore Phocas (963–969)*, Paris: 199–315.

Cheynet, J.C. 1986b. 'Notes arabo-byzantines'. In: V. Kremmydas, C. Maltezou and N.M. Panagiotakis (eds.), *Αφιέρωμα στον Νίκο Σβορώνο*, vol. 1, Rethymno: 145–52.

Cheynet, J.C. 1991. 'Fortune et puissance de l'aristocratie (Xe–XIIe siècle)'. In: V. Kravari, J. Lefort and C. Morrisson (eds.), *Hommes et richesses dans l'empire byzantine II*, Paris: 199–214.

Cheynet, J.C. 1995. 'Les effectifs de l'armée byzantine aux Xe–XIIe siècles'. *Cahiers de civilization médiévale* 38: 319–35.

Cheynet, J.C. 2001. 'La conception militaire de la frontière orientale (IXᵉ–XIIIᵉ siècle)'. In: A. Eastmond (ed.), *Eastern Approaches to Byzantium: Papers from the Thirty-Third Spring Symposium of Byzantine Studies, University of Warwick, Coventry, March 1999*, Aldershot and Burlington, VT: 57–69.

Cheynet, J.C. 2006. 'The Byzantine aristocracy'. In: J.C. Cheynet (ed.), *The Byzantine Aristocracy and Its Military Function*, trans. J. Lefort, Farnham and Burlington, VT: 1–43.

Cheynet, J.C. 2014. 'La pensée stratégique byzantine'. In: J. Baechler and J.V. Holeindre (eds.), *Penseurs de la stratégie*, Paris: 45–58.

Cheynet, J.C. and Vannier, J.F. 2003. 'Les Argyroi'. *ZRVI* 40: 57–90.

Christides, V. 1969. 'Arabs as "Barbaroi" before the rise of Islam'. *Balkan Studies* 10: 317–24.

Christides, V. 1984. 'Naval warfare in the eastern Mediterranean (6th–14th centuries): An Arabic translation of Leo VI's *Naumachica*'. *Graeco-Arabica* 3: 137–43.

Christides, V. 1995. 'Ibn al – Manqalī (Manglī) and Leo VI: New evidence on Arab-Byzantine ship construction and naval warfare'. *BS* 56: 83–96.

Codoñer, J.S. 2014a. *The Emperor Theophilos and the East, 829–842: Court and Frontier in Byzantium during the Last Phase of Iconoclasm*, Farnham and Burlington, VT.

Codoñer, J.S. 2014b. 'Towards a vocabulary for rewriting in Byzantium'. In: J.S. Codoñer and I. Peréz-Martín (eds.), *Textual Transmission in Byzantium: Between Textual Criticism and Quellenforschung*, Turnhout: 61–90.

Cohn, I. 1900. 'Bemerkugen zu den Konstantinischen Sammelwerken'. *BZ* 9: 155–60.

Cooper, J.E. and Decker, M.J. 2012. *Life and society in Byzantine Cappadocia*, Basingstoke and New York.

Cosentino, S. 2000. 'The Syrianos's "Strategikon": A 9th century source?'. *Bizantinistica* 2: 243–80.

Croke, B. 2006. 'Tradition and originality in Photius' historical reading'. In: J. Burke et al. (eds.), *Byzantine Narrative: Papers in Honour of Roger Scott*, Melbourne: 59–70.

Cutler, A. 1995. 'Originality as a cultural phenomenon'. In: A.R. Littlewood (ed.), *Originality in Byzantine Literature, Art and Music: A Collection of Essays*, Oxford: 203–16.

Da Costa-Louillet, G. 1954. 'Saints de Constantinople aux VIIIe, IXe, Xe siècles'. *Byzantion* 24: 453–511.

Dagron, G. 1983. 'Byzance et le modèle islamique au Xe siècle. À propos de Constitutions tactiques de l'empereur Léon VI'. In: *Comptes rendus des séances de l'Académie des Inscriptions et Belles-Lettres*, Paris: 219–43.

Dagron, G. 1984. *Constantinople imaginaire. Études sur le recueil des "Patria"*, Paris.

Dagron, G. and Mihăescu, H. 1986. 'Commentary'. In: *Le traité sur la guérrilla (De velitatione) de l'empereur Nicéphore Phocas (963–969)*, Paris: 139–287.

Dain, A. 1931. 'Les cinq adaptations byzantines des *Stratagèmes* de Polyen'. *Revue des Études Anciennes* 33: 321–45.

Dain, A. 1933. *La tradition du texte d'Héron de Byzance*, Paris.

Dain, A. 1937. *La 'Tactique' de Nicéphore Ouranos*, Paris.

Dain, A. 1938. 'Introduction'. In: *Sylloge tacticorum, quae olim "Inedita Leonis tactica" dicebatur*, Paris: 7–10.

Dain, A. 1939. *Le Corpus Perditum*, Paris.

Dain, A. 1940a. 'Introduction'. In: 'Mémorandum inédit sur la défense des places'. *Revue des Études Grecques* 53: 123–36.

Dain, A. 1940b. *La collection florentine des tacticiens grecs*, Paris.

Dain, A. 1946. *Histoire du texte d'Élien le tacticien des origines à la fin du moyen âge*, Paris.

Dain, A. 1953. 'L'encyclopédisme de Constantin Porphyrogénète'. *Lettres d'humanité* 12: 64–81.

Dain, A. and Foucault, J.A. 1967. 'Les strategists byzantins'. *TM* 2: 317–92.

Darrouzès, J. and Westerink, L.G. 1978. 'Introduction'. In: *Théodore Daphnopatès Correspondance*, Paris: 1–27.

Dawson, T. 1999. 'Kremasmata, kabadion, klibanion: Some aspects of Middle Byzantine equipment reconsidered'. *BMGS* 22: 38–50.

Dawson, T. 2001–2002. 'Klivanion revisited: An evolutionary typology and catalogue of Middle Byzantine lamellar styles'. *Journal of Roman Military Equipment Studies* 12/13: 89–95.

Dawson, T. 2002. 'Suntagma Hoplon: The equipment of regular Byzantine troops, c.950 to c.1204'. In: D. Nicolle (ed.), *A Companion to Medieval Arms and Armour*, London: 81–90.

Dawson, T. 2007. 'Fit for the task: Equipment sizes and the transmission of military lore, sixth to tenth centuries'. *BMGS* 31: 1–12.

Decker, M.J. 2013. *The Byzantine Art of War*, Yardley, PA.

Dennis, G.T. 1981. 'Introduction'. In: *Das Strategikon des Maurikios*, Vienna: 13–42.

Dennis, G.T. 1984. 'Introduction'. In: *Maurice's Strategikon: Handbook of Byzantine Military Strategy*, Philadelphia: vii–xxi.

Dennis, G.T. 1985. 'Introduction'. In: *Three Byzantine Military Treatises*, Washington, DC: 1–7, 137–41, 241–4.

Dennis, G.T. 1988. 'The Byzantines as revealed in their letters'. In: J. Duffy and J. Peradotto (eds.), *Gonimos: Neoplatonic and Byzantine Studies Presented to Leendert G. Westerink at 75*, Buffalo, NY: 155–65.

Dennis, G.T. 1993. 'Religious services in the Byzantine army'. In: E. Carr et al. (eds.), *Eulogēma: Studies in honor of Robert Taft, S.J.*, Rome: 107–17.

Dennis, G.T. 1997. 'The Byzantines in battle'. In: K. Tsiknakis (ed.), *Το εμπόλεμο Βυζάντιο (9ος–12ος αι.)*, Athens: 165–78.

Dennis, G.T. 1998. 'Byzantine heavy artillery: The Helepolis'. *GRBS* 39: 99–115.

Dennis, G.T. 2014. 'Introduction'. *The Taktika of Leo VI*, Washington, DC: ix–xiv.

Diethart, J.M. and Dintsis, P. 1984. 'Die Leontoklibanarier: Versuch einer archäologisch-papyrologisch Zusammenschau'. *BYZANTIOΣ: Festschrift für H. Hunger zum 70. Geburtstag*, Vienna.

Diller, A. 1950. 'Julian of Ascalon on Strabo and the Stade'. *Classical Philology* 45: 22–25.

Dixon, K.R. and Southern, P. 1992. *The Roman Cavalry: From the First to the Third Century AD*, London.

Dujčev, I. 1978. 'On the treaty of 927 with the Bulgarians'. *DOP* 32: 217–95.

Eadie, J.W. 1967. 'The development of Roman mailed cavalry'. *Journal of Roman Studies* 57: 161–73.

Eickhoff, E. 1966. *Seekrieg und Seepolitik zwischen Islam und Abendland: Das Mittelmeer unter byzantinischer und arabischer Hegemonie (650–1040)*, Berlin.

El Cheikh, N.M. 2004. *Byzantium Viewed by the Arabs*, Cambridge, MA and London.

Fear, A. 2008. 'War and society'. In: P. Sabin, H.V. Wees and M. Whitby (eds.), *The Cambridge History of Greek and Roman Warfare Volume II: Rome from the Late Republic to the Late Empire*, Cambridge: 424–58.

Featherstone, M. 2004. 'Further remarks on the De Cerimoniis'. *BZ* 91: 113–21.

Featherstone, M. 2011. 'Theophanes Continuatus VI and *De Cerimoniis* I, 96'. *BZ* 104: 106–19.

Featherstone, M. 2012. 'Theophanes Continuatus: A history for the palace'. In: P. Odorico (ed.), *La face cachée de la littérature byzantine: Le texte en tant que message immédiat*, Paris: 123–35.

Featherstone, M. 2013. '*De Cerimoniis*: The revival of antiquity in the Great Palace and the Macedonian Renaissance'. In: N. Necipoğlu, A. Ödekan and E. Akyürek (eds.), *The Byzantine Court: Source of Power and Culture (2nd International Sevgi Gönül Byzantine Studies Symposium, Istanbul 21–23 June 2010)*, Istanbul: 137–44.

Featherstone, M. 2014. 'Basileios Nothos as compiler: The *De Cerimoniis* and Theophanes Continuatus'. In: J.S. Codoñer and I.P. Martín (eds.), *Textual Transmission in Byzantium: Between Textual Criticism and Quellenforschung*, Turnhout: 353–72.

Flusin, B. 2001. 'L'empereur hagiographe. Remarques sur le rôle des premiers empereurs macédoniens dans le culte des saints'. In: P. Guran (ed.), *L'empereur hagiographe: Culte des saints et monarchie byzantine et post-byzantine, Actes des colloques internationaux "L'empereur hagiographe" (13–14 mars 2000) et "Reliques et miracles" (1–2 novembre 2000) tenus au New Europe College*, Bucharest: 29–54.

Flusin, B. 2002. 'Les Excerpta constantiniens: logique d'une anti-histoire'. In: S. Pittia (ed.), *Fragments d'historiens grecs: autour de Denys d'Halicarnasse*, Rome: 537–59.

Forsyth, J.H. 1977. *The Byzantine-Arab Chronicle (938–1034) of Yaḥyā b. Saʿīd al-Anṭākī*, PhD thesis, University of Michigan, Ann Arbor, MI.

Foucault, J.A. 1973. 'Douze chapitres inédits de la "Tactique" de Nicéphore Ouranos'. *TM* 5: 281–312.

France, J. 1994. *Victory in the East: A Military History of the First Crusade*, Cambridge.

Franklin, S. and Shepard, J. 1996. *The Emergence of Rus 750–1200*, London and New York.

Frankopan, P. 2012. *The First Crusade: The Call from the East*, Cambridge, MA.

Frankopan, P. 2013. 'Turning Latin into Greek: Anna Komnene and the *Gesta Roberti Wiscardi*'. *Journal of Medieval History* 39: 80–99.

Frankopan, P. 2014. 'Understanding the Greek sources for the First Crusade'. In: M. Bull and D. Kempf (eds.), *Writing the Early Crusades: Text, Transmission and Memory*, Woodbridge: 38–52.

Fryde, E.B. 1983. *Humanism and Renaissance Historiography*, London.

Fryde, E.B. 1996. *Greek Manuscripts in the Private Library of the Medici, 1469–1510*, vol. 2, Aberystwyth.

Gamber, O. 1968. 'Kataphrakten, Clibanarier, Normannenreiter'. *Jahrbuch der Kunsthistorischen Sammlungen in Wien* 64: 7–44.

Garrood, W. 2013. 'The illusion of continuity: Nikephoros Phokas, John Tzimiskes and the eastern border'. *BMGS* 37: 20–34.

Geiger, J. 1992. 'Julian of Ascalon'. *Journal of Hellenic Studies* 112: 31–43.

Gerlach, J. 2008. *Gnomica Democritea: Studien zur gnomologischen Überlieferung der Ethik Demokrits und zum Corpus Parisinum mit einer Edition der Democritea des Corpus Parisinum*, Wiesbaden.

Gordon, M. 2001. 'The commanders of the Samarran Turkish military: The shaping of a third/ninth-century imperial elite'. In: C.F. Robinson (ed.), *A Medieval Islamic City Reconsidered: An Interdisciplinary Approach to Samarra*, Oxford: 119–40.

Grégoire, H. 1934. 'Manuel et Théophobe ou le concurrence de deux monastères'. *Byzantion* 9: 183–204.

Grégoire, H. 1938. 'Saint Théodore le Stratélate et les Russes d'Igor'. *Byzantion* 13: 291–300.

Grégoire, H. and Orgels, P. 1954. 'L'invasion hongroise dans la "Vie de saint Basile le jeune"'. *Byzantion* 24: 147–54.

Grigoriou-Ioannidou, M. 1983. 'Η βυζαντινοβουλγαρική σύγκρουση στους Κατασύρτες (917)'. *Επιστημονική Επετηρίδα Φιλοσοφικής Σχολής ΑΠΘ* 21: 123–48.

Grigoriou-Ioannidou, M. 1993. 'Θέματα et Τάγματα, Un problème de l' institution des thèmes pendant les Xe et XIe siècles'. *BF* 19: 35–41.

Grigoriou-Ioannidou, M. 1999. 'Οι Ούγγροι και οι επιδρομές τους στον δυτικό-ευρωπαϊκό και στον βυζαντινό χώρο (τέλη 9ου–10ος αι.)'. *Βυζαντινά* 20: 65–135.

Grosdidier de Matons, J. 1973. 'Les constitutions tactiques et la damnatio memoriae de l'empereur Alexandre'. *TM* 5: 229–42.

Grotowski, P.L. 2010. *Arms and Armour of the Warrior Saints, Tradition and Innovation in Byzantine Iconography (843–1261)*, trans. R. Brzezinski, Leiden and Boston.

Guilland, R. 1950. 'Études sur l'histoire administrative de Byzance: Le Domestique des Scholes'. *REB* 8: 5–63.

Guilland, R. 1967. *Recherches sur les institutions byzantines*, vol. 1, Berlin and Amsterdam.

Guscin, M. 2009. *The Image of Edessa*, Leiden and Boston.

Gyftopoulou, S. 2013. 'Historical information gathered from *Mauricii Strategikon*'. *Βυζαντινά Σύμμεικτα* 23: 59–89.

Haase, F. 1847. *De militarium scriptorum Graecorum et Latinorum omnium editione instituenda narratio*, Berlin.

Hakim, S.B. 2001. 'Julian of Ascalon's treatise of construction and design rules from sixth-century Palestine'. *Journal of the Society of Architectural Historians* 60: 4–25.

Haldon, J.F. 1975. 'Some aspects of Byzantine military technology from the sixth to the tenth centuries'. *BMGS* 1: 11–46.

Haldon, J.F. 1979. *Recruitment and Conscription in the Byzantine Army c.550–950: A Study of the Stratiotika Ktemata*, Vienna.

Haldon, J.F. 1984. *Byzantine Praetorians: An Administrative, Institutional and Social Survey of the Opsikion and Tagmata, c.580–900*. Bonn.

Haldon, J.F. 1990a. 'Introduction–Commentary'. In: *Constantine Porphyrogenitus: Three Treatises on Imperial Military Expeditions*, Vienna: 35–78, 155–293.

Haldon, J.F. 1990b. *Byzantium in the Seventh Century: The Transformation of a Culture*, revised edition, Cambridge.

Haldon, J.F. 1993. 'Military service, military lands, and the status of soldiers: Current problems and interpretations'. *DOP* 47: 1–67.

Haldon, J.F. 1999. *Warfare, State and Society in the Byzantine World, 565–1204*, London.

Haldon, J.F. 2000. 'Theory and practice in the 10th century military administration'. *TM* 13: 201–352.

Haldon, J.F. 2002. 'Some aspects of Early Byzantine arms and armour'. In: D. Nicolle (ed.), *A Companion to Medieval Arms and Armour*, London: 65–79.

Haldon, J.F. 2006. 'Greek Fire revisited: Recent and current research'. In: E. Jeffreys (ed.), *Byzantine Style, Religion and Civilization: In Honour of Sir Steven Runciman*, Cambridge: 290–325.

Haldon, J.F. 2008. 'Provincial elites, central authorities in the Byzantine state'. In: B. Forsén and G. Salmeri (eds.), *The Province Strikes Back, Imperial Dynamics in the Eastern Mediterranean*, Helsinki: 157–85.

Haldon, J.F. 2009. 'Social elites, wealth and power'. In: J.F. Haldon (ed.), *A Social History of Byzantium*, Chichester and Malden, MA: 168–211.

Haldon, J.F. 2013. 'Information and war: Some comments on defensive strategy and information in the Middle Byzantine period (ca. A.D. 660–1025)'. In: A. Sarantis and N. Christie (eds.), *War and Warfare in Late Antiquity*, vol. 2, Leiden and Boston: 373–93.

Haldon, J.F. 2014. *A Critical Commentary on the Taktika of Leo VI*, Washington, DC.

Haldon, J.F. 2016. 'Introduction'. In: *A Tale of Two Saints: The Martyrdoms and Miracles of Saints Theodore "the Recruit" and "the General"*, Liverpool: 1–19.

Haldon, J.F. and Kennedy, H. 1980. 'The Arab-Byzantine frontier in the eighth and ninth centuries: Military organisation and society in the borderlands'. *ZRVI* 19: 79–116.

Harris, J. 1995. *Greek Emigres in the West 1400–1520*, Camberley.

Harris, J. 2014. *Byzantium and the Crusades*, 2nd edition, London and New York.

Hinterberger, M. 2010. 'Envy and nemesis in the *Vita Basilii* and *Leo the Deacon*: Literary mimesis or something more?'. In: R. Macrides (ed.), *History as Literature in Byzantium: Papers from the Fortieth Spring Symposium of Byzantine Studies, University of Birmingham, March 2007*, Farnham and Burlington, VT: 187–206.

Hoffmann, L. 2007. 'Geschichtsschreibung oder Rhetorik? Zum *logos parakletikos* bei Leon Diakonos'. In: M. Grünbart (ed.), *Rhetorische Kultur in Spätantike und Mittelalter/ Rhetorical Culture in Late Antiquity and the Middle Ages*, Berlin: 105–39.

Holmes, C. 2001. 'Book Review: *Siegecraft: Two Tenth-Century Instructional Manuals by "Heron of Byzantium"'*. *War in History* 8: 479–81.

Holmes, C. 2002. 'Byzantium's eastern frontier in the tenth and eleventh centuries'. In: D. Abulafia and N. Berend (eds.), *Medieval Frontiers: Concepts and Practices*, Aldershot and Burlington, VT: 83–104.

Holmes, C. 2005. *Basil II and the Governance of Empire*, Oxford.

Holmes, C. 2008. 'Treaties between Byzantium and the Islamic world'. In: P. de Souza and J. France (eds.), *War and Peace in Ancient and Medieval History*, Cambridge: 141–57.

Holmes, C. 2010. 'Byzantine political culture and compilation literature in the tenth and eleventh centuries: Some preliminary inquiries'. *DOP* 64: 55–80.

Howard-Johnston, J.D. 1983. 'Byzantine Anzitene'. In: S. Mitchell (ed.), *Armies and Frontiers in Roman and Byzantine Anatolia, Proceedings of a Colloquium Held at University College, Swansea, in April 1981*, Oxford: 239–90.

Howard-Johnston, J.D. 1994. 'The official history of Heraclius' Persian campaigns'. In: E. Dąbrowa (ed.), *The Roman and Byzantine Army in the East: Proceedings of a Colloquium Held at the Jagiellonian University, Kraków in September 1992*, Kraków: 57–87.

Howard-Johnston, J.D. 1995. 'Crown lands and the defence of imperial authority in the tenth and eleventh centuries'. *BF* 21: 75–100.

Howard-Johnston, J.D. 2000. 'The *De Administrando Imperio*: A re-examination of the text and a re-evaluation of its evidence about the Rus'. In: M. Kazanski, A. Nercessian and C. Zuckerman (eds.), *Les centres proto-urbains russes entre Scandinavie, Byzance et Orient*, Paris: 301–36.

Howard-Johnston, J.D. 2010. *Witnesses to a World Crisis: Historians and Histories of the Middle East in the Seventh Century*, Oxford.

Hunger, H. 1969–1970. 'On the imitation (ΜΙΜΗΣΙΣ) of Antiquity in Byzantine literature'. *DOP* 23/24: 15–38.

Hunger, H. 1978. *Die hochsprachliche profane Literatur der Byzantiner*, 2 vols., Munich.

Huxley, G.L. 1975. 'A List of ἄπληκτα'. *GRBS* 16: 87–93.

Huxley, G.L. 1980. 'The scholarship of Constantine Porphyrogenitus'. *Proceeding of the Royal Irish Academy: Section C: Archaeology, Celtic Studies, History, Linguistics, Literature* 80C: 29–40.

Irigoin, J. 1959. 'Pour une étude des centres de copie byzantins'. *Scriptorium* 13: 177–209.

Irigoin, J. 1997. 'Lascaris Rhyndacenus (Janus) (1445–1535)'. In: C. Nativel (ed.), *Centuriae Latinae, Cent une figures humanistes de la Renaissance aux Lumières offertes à Jacques Chomarat*, Geneva: 485–91.

Jackson, D.F. 1998. 'A new look at an old book list'. *Studi italiani di filologia classica* 16: 83–108.

Jackson, D.F. 2003. 'Janus Lascaris on the island of Corfu in A.D. 1491'. *Scriptorium* 57: 137–9.

Jenkins, R.J.H. 1954. 'The classical background of the *Scriptores Post Theophanem*'. *DOP* 8: 13–30.

Jenkins, R.J.H. 1966. 'The peace with Bulgaria (927) celebrated by Theodore Daphnopates'. In: R. Jenkins (repr.), *Studies on Byzantine History of the 9th and 10th Centuries*, London: 287–303.

Jones, A.H.M. 1964. *The Later Roman Empire 284–602: A Social, Economic, and Administrative Survey*, Oxford.

Kaegi, W.E.J. 1964a. 'The contribution of archery to the Turkish conquest of Anatolia'. *Speculum* 39: 96–108.

Kaegi, W.E.J. 1964b. 'The Emperor Julian's assessment of the significance and function of history'. *Proceedings of the American Philosophical Society* 108: 29–38.

Kaegi, W.E.J. 1981. 'Constantine's and Julian's strategies of strategic surprise against the Persians'. *Athenaeum* 69: 209–13.

Kaegi, W.E.J. 1983. *Some Thoughts on Byzantine Military Strategy*, Brookline, MA.

Kaegi, W.E.J. 1990. 'Procopius the military historian'. *BF* 15: 53–85.

Kaegi, W.E.J. 1992. *Byzantium and the Early Islamic Conquests*, Cambridge.

Kaegi, W.E.J. 2003. *Heraclius, Emperor of Byzantium*, Cambridge.

Kaldellis, A. 1998. 'Introduction and commentary'. In: *Genesios: On the Reigns of the Emperors*, Canberra: 45–7.

Kaldellis, A. 2007. *Hellenism in Byzantium: The Transformations of Greek Identity and the Reception of the Classical Tradition*, Cambridge.

Kaldellis, A. 2012. 'Byzantine historical writing, 500–920'. In: S. Foot and C.F. Robinson (eds.), *The Oxford History of Historical Writing Volume 2: 400–1400*, Oxford: 201–17.

Kaldellis, A. 2013. 'The original source for Tzimiskes' Balkan campaign (971 AD) and the Emperor's classicizing propaganda'. *BMGS* 31: 35–52.

Kaldellis, A. 2014. 'Did Ioannes I Tzimiskes campaign in the east in 974?'. *Byzantion* 84: 235–40.

Kaldellis, A. 2015a. 'The Byzantine conquest of Crete (961 AD), Procopius' *Vandal War*, and the continuator of the *Chronicle* of Symeon'. *BMGS* 39: 302–11.

Kaldellis, A. 2015b. *Byzantine Readings of Ancient Historians*, Abingdon and New York.

Kaldellis, A. 2017. *Streams of Gold, Rivers of Blood: The Rise and Fall of Byzantium, 955 A.D. to the First Crusade*, New York.

Karantabias, M.A. 2005–2006. 'The crucial development of heavy cavalry under Herakleios and his usage of Steppe Nomad tactics'. *Hirundo* 4: 28–41.

Karapli, K.G. 2010. *Κατευόδωσις στρατού: Η οργάνωση και η ψυχολογική προετοιμασία του βυζαντινού στρατού πριν από τον πόλεμο (610–1081)*, Athens.

Karayannopulos, J. 1959. *Die Entstehung der byzantinischen Themenordung*, Munich.

Karlin-Hayter, P. 1969. 'The Emperor Alexander's bad name'. *Speculum* 44: 586–96.

Karpozilos, A. 2002–2009. *Βυζαντινοί ιστορικοί και χρονογράφοι*, 2–3 vols., Athens.

Kazhdan, A.P. 1978. 'Some questions addressed to the scholars who believe in the authenticity of Kaminiates' "Capture of Thessalonica"'. *BZ* 71: 301–14.

Kazhdan, A.P. 1995. 'Innovation in Byzantium'. In: A.R. Littlewood (ed.), *Originality in Byzantine Literature, Art and Music: A Collection of Essays*, Oxford: 1–14.

Kazhdan, A.P. 2006. *A History of Byzantine Literature (850–1000)*, ed. C. Angelidi, Athens.

Kazhdan, A.P. and Epstein, A.W. 1985. *Change in Byzantine Culture in the Eleventh and Twelfth Centuries*, Berkeley, CA, Los Angeles and London.

Kazhdan, A.P., Talbot, A.-M., Cutler, A., Gregory, T.E. and Ševčenko, N.P. (eds.). 1991. *Oxford Dictionary of Byzantium*, 3 vols., New York and Oxford.

Kennedy, G.A. 2003. *Progymnasmata: Greek Textbooks of Prose Composition and Rhetoric*, Atlanta, GA.

Kennedy, H. 2001. *The Armies of the Caliphs, Military and Society in the Early Islamic State*, London and New York.

Kennedy, H. 2004. *The Prophet and the Age of the Caliphates*, 2nd edition, Harlow.

Kiapidou, E.S. 2010. *Η Σύνοψη Ιστοριών του Ιωάννη Σκυλίτζη και οι πηγές της (811–1057)*, Athens.

Knös, B. 1945. *Un ambassadeur de l'Hellénisme Janus Lascaris et la tradition Greco-Byzantine dans l'humanisme français*, Stockholm and Paris.

Köchly, H. 1854. 'Introduction'. In: *Index lectionum in literarum universitate Turicensi habendarum*, 2 vols., Zurich.

Köchly, H. and Rüstow, W. 1855. *Griechische Kriegsschriftsteller*, Leipzig.

Kolia-Dermitzaki, A. 1989. 'Η ιδέα του "ιερού πολέμου" στο Βυζάντιο κατά τον 10° αιώνα: Η μαρτυρία των τακτικών και των δημηγοριών'. In: A. Markopoulos (ed.), *Κωνσταντίνος Ζ΄ ο Πορφυρογέννητος και η εποχή του. Β΄ Διεθνής Βυζαντινολογική Συνάντηση, Δελφοί, 22–26 Ιουλίου 1987*, Athens: 39–55.

Kolia-Dermitzaki, A. 1997. 'Το εμπόλεμο Βυζάντιο στις ομιλίες και τις επιστολές του 10ου και 11ου αιώνα. Μια ιδεολογική προσέγγιση'. In: K. Tsinakis (ed.), *Το εμπόλεμο Βυζάντιο (9ος–12ος αι.)*, Athens: 213–38.

Kolia-Dermitzaki, A. 2000. 'Some remarks on the fate of prisoners in Byzantium (9th–10th centuries)'. In: G. Cipollone (ed.), *La liberazione dei "captivi" tra Cristianità e Islam. Otre la crociata e il Ǧihād: Tolleranza e servizio umanitario*, Vatican City: 583–620.

Kolias, T.G. 1984a. 'The Taktika of Leo VI and the Arabs'. *Graeco-Arabica* 3: 129–35.

KoliasKolias, T.G. 1984b. 'Eßgewohnheiten und Verpflegung im byzantinischen Heer'. In: W. Hörandner et al. (eds.), *Byzantios: Festschrift für Herbert Hunger zum 70. Geburtstag*, Vienna: 193–202.

Kolias, T.G. 1988. *Byzantinische Waffen, Ein Beitrag zur byzantinischen Waffenkunde von den Anfängen bis zur lateinischen Eroberung*, Vienna.

Kolias, T.G. 1989. 'Τα όπλα στην βυζαντινή κοινωνία'. In: C. Angelidi (ed.), *Η καθημερινή ζωή στο Βυζάντιο. Τομές και συνέχειες στην ελληνιστική και ρωμαϊκή παράδοση, πρακτικά του Α διεθνούς συμποσίου*, Athens: 463–76.

Kolias, T.G. 1993a. *Νικηφόρος [Β΄] Φωκάς (963–969): Ο στρατηγός αυτοκράτωρ και το μεταρρυθμιστικό του έργο*, Athens.

Kolias, T.G. 1993b. 'Tradition und Erneuerung im Frühbyzantinischen Reich am Beispiel der militärischen Sprache und Terminologie'. In: F. Vallet and M. Kazanski (eds.), *L'armée romaine et les barbares du IIIᵉ au VIIᵉ siècle*, Rouen: 39–44.

Kolias, T.G. 1994. *Η θέση του στρατιωτικού στην βυζαντινή κοινωνία*, Athens.

Kolias, T.G. 1995. 'Kriegsgefangene, Sklavenhandel und die Privilegien der Soldaten. Die Aussage der Novelle von Ioannes Tzimiskes'. *BS* 56: 129–35.

Kolias, T.G. 1997. 'Η πολεμική τακτική των Βυζαντινών: Θεωρία και πράξη'. In: K. Tsinakis (ed.), *Το εμπόλεμο Βυζάντιο (9ος–12ος αι.)*, Athens: 153–64.

Kolias, T.G. 2007. 'Zur Gerichtsbarkeit im byzantinischen Heer'. In: K. Belke et al. (eds.), *Byzantina Mediterranea: Festschrift für J. Koder zum 65 Geburtstag*, Vienna, Cologne and Weimar: 319–25.

Korres, Th. 1995. *"Υγρόν Πυρ". Ένα όπλο της βυζαντινής ναυτικής τακτικής*, Thessaloniki.

Korzenszky, E. 1931. 'Introduction'. In: *Leges poenales militares e codice Laurentiano LXXV6*, Budapest: 155–63.

Koutava-Delivoria, B. 1991. *Ο γεωγραφικός κόσμος Κωνσταντίνου του Πορφυρογεννήτου*, vol. 1, Athens.

Koutava-Delivoria, B. 2002. 'La contribution de Constantin porphyrogénète à la composition des Geoponica'. *Byzantion* 72: 365–80.

Koutrakou, N. 1995. 'Diplomacy and espionage: Their role in Byzantine foreign relations, 8th–10th centuries'. *Graeco-Arabica* 6: 125–44.

Krallis, D. 2012. *Michael Attaleiates and the Politics of Decline in Eleventh-Century Byzantium*, Tempe, AZ.

Krentz, P. and Wheeler, E.L. 1994. 'Introduction'. In: *Polyaenus, Stratagems of War*, Chicago: vi–xxiv.

Krsmanović, B. 2008. *The Byzantine Province in Change (On the Threshold between the 10th and the 11th Century)*, Belgrade and Athens.

Krumbacher, K. 1897. *Geschichte der byzantinischen Literatur von Justinian bis zum Ende des Oströmischen Reiches (527–1453)*, Munich.

Kühn, H.J. 1991. *Die byzantinische Armee im 10. und 11. Jahrhundert, Studien zur Organisation der Tagmata*, Vienna.

Kyriakis, E.K. 1993. *Βυζάντιο και Βούλγαροι (7ος–10ος αι.): Συμβολή στην εξωτερική πολιτική του Βυζαντίου*, Athens.

Kyriakidis, S. 2009. 'The division of booty in Late Byzantium (1204–1453)'. *JÖB* 59: 163–75.

Kyriakidis, S. 2016a. 'Accounts of single combat in Byzantine historiography in the 10th to 14th centuries'. *Acta Classica* 59: 114–36.

Kyriakidis, S. 2016b. 'The battle exhortation in Byzantine historiography (10th–12th centuries)'. *Erytheia* 37: 19–36.

Lechner, K. 1955. *Hellenen und Barbaren im Weltbild der Byzantiner, die alten Bezeichnungen als Ausdruck eines neuen Kulturbewußtseins*, Munich.

Lemerle, P. 1971. *Le premier humanisme byzantin: Notes et remarques sur enseignement et culture à Byzance des origines au Xe siècle*, Paris.

Lemerle, P. 1972. 'Séance de clôture'. In: *Πρακτικά του Πρώτου Διεθνούς Κυπρολογικού Συνεδρίου*, vol. 2, Nicosia: 151–6.

Lendon, J.E. 2005. *Soldiers and Ghosts: A History of Battle in Classical Antiquity*, New Haven, CT and London.

Lev, Y. 1995. 'The Fatimids and Byzantium, 10th–12th centuries'. *Graeco-Arabica* 6: 190–208.

Lilie, R.-J. 1976. *Die byzantinische Reaktion auf die Ausbreitung der Araber: Studien zur Strukturwandlung des byzantinischen Staates im 7. Und 8. Jhd*, Munich.

Lilie, R.-J. 1993. 'Die Zentralbürokratie die Provinzen zwischen dem 10. und dem 12. Jahrhundert. Anspruch und Realität'. *BF* 19: 65–75.

Lilie, R.-J. 2014. 'Reality and invention: Reflections on Byzantine historiography'. *DOP* 68: 157–210.

Ljubarskij, J. 1993. 'Nikephoros Phokas in Byzantine historical writings: Trace of the secular biography in Byzantium'. *BS* 54: 245–53.

Loud, G.A. 1991. 'Anna Komnena and her sources for the Normans of southern Italy'. In: I. Wood and G.A. Loud (eds.), *Church and Chronicle in the Middle Ages, Essays Presented to John Taylor*, London: 41–57.

Lounghis, T.K. 1997. 'Επιθεώρηση ενόπλων δυνάμεων πριν από εκστρατεία'. In: K. Tsiknakis (ed.), *Το εμπόλεμο Βυζάντιο (9ος – 12ος αι.)*, Athens: 93–110.

Luttwak, E.N. 2009. *The Grand Strategy of the Byzantine Empire*, Cambridge, MA and London.

Macrides, R. 2000. 'The pen and the sword: Who wrote the *Alexiad*?'. In: T. Gouma-Peterson (ed.), *Anna Komnene and Her Times*, New York and London: 63–82.

Madgearu, A. 2013. *Byzantine Military Organization on the Danube, 10th–12th Centuries*, Leiden and Boston.

Magdalino, P. 1987. 'Observations on the Nea Ekklesia of Basil I'. *JÖB* 37: 51–64.

Magdalino, P. 1993. *The Empire of Manuel I Komnenos 1143–1180*, Cambridge.

Magdalino, P. 1999. 'What we heard in the lives of the saints we have seen with our own eyes': The holy man as literary text in tenth-century Constantinople'. In: J. Howard-Johnston and P.A. Hayward (eds.), *The Cult of Saints in Late Antiquity and the Middle Ages*, Oxford: 83–112.

Magdalino, P. 2000. 'The pen of the aunt: Echoes of the mid-twelfth century in the *Alexiad*'. In: T. Gouma-Peterson (ed.), *Anna Komnene and Her Times*, New York and London: 15–44.

Magdalino, P. 2009. 'Court society and aristocracy'. In: J.F. Haldon (ed.), *A Social History of Byzantium*, Chichester and Malden, MA: 212–32.

Magdalino, P. 2013. 'Knowledge in authority and authorised history: The imperial intellectual programme of Leo VI and Constantine VII'. In: P. Armstrong (ed.), *Authority in Byzantium*, Farnham and Burlington, VT: 187–209.

Makrypoulias, C.G. 2000. 'Byzantine expeditions against the Emirate of Crete c. 825–949'. *Graeco-Arabica* 7–8: 347–62.

Makrypoulias, C.G. 2012. 'Civilians as combatants in Byzantium: Ideological versus practical considerations'. In: J. Koder and I. Stouraitis (eds.), *Byzantine War Ideology between Roman Imperial Concept and Christian Religion: Akten des Internationalen Symposiums (Wien, 19.–21. Mai 2011)*, Vienna: 109–20.

Makrypoulias, C.G. 2013. 'Η μελέτη της βυζαντινής πολεμικής τεχνολογίας: Προβλήματα μεθοδολογίας'. In: E. Mergupe-Sabaidu et al. (eds.), *Επιστήμη και τεχνολογία. Ιστορικές και ιστοριογραφικές μελέτες*, Athens: 31–44.

Malamut, E. 1988. *Les îles de l'empire byzantin VIIIe–XIIe siècles*, 2 vols., Paris.

Mango, C. 1975. *Byzantine Literature as a Distorting Mirror: An Inaugural Lecture Delivered before the University of Oxford on 21 May 1974*, Oxford.

Mango, C. 1980. *Byzantium: The Empire of New Rome*, London.

Mango, C. 1982. 'The Life of Saint Andrew the Fool reconsidered'. *RSBS* 2: 297–313.

Mango, C. 2011. 'Introduction'. In: I. Ševčenko (ed.), *Chronographiae quae Theophanis Continuati nomine fertur liber quo Vita Basilii imperatoris amplectitur*, Berlin: 3–13.

Markesinis, B. 2000. 'Janos Lascaris, la bibliothèque d'Avramis à Corfou et le Paris. gr. 854'. *Scriptorium* 54: 302–6.

Markopoulos, A. 1994. 'Constantine the Great in Macedonian historiography: Models and approaches'. In: P. Magdalino (ed.), *New Constantines: The Rhythm of Imperial Renewal in Byzantium, 4th–13th Centuries*, Aldershot and Burlington, VT: 159–70.

Markopoulos, A. 1998. 'Autour des Chapitres Parénétiques de Basile Ier'. In: M. Balard et al. (eds.), *Εὐψυχία: Mélanges offerts à Hélène Ahrweiler*, Paris: 469–79.

Markopoulos, A. 2003. 'Byzantine history writing at the end of the first millennium'. In: P. Magdalino (ed.), *Byzantium in the Year 1000*, Leiden and Boston: 183–97.

Markopoulos, A. 2006. 'Η ιστοριογραφία των δυνατών κατά τη μεσοβυζαντινή περίοδο. Ο Ιωάννης Κουρκούας στην ιστορική συγγραφή του πρωτοσπαθάριου και κριτή Μανουήλ'. In: G.K. Promponas and P. Valavanis (eds.), *Ευεργεσίη: Τόμος χαριστήριος στον Παναγιώτη Ι. Κοντό*, vol. 1, Athens: 397–405.

Markopoulos, A. 2009. 'From narrative historiography to historical biography: New trends in Byzantine historical writing in the 10th–11th Centuries'. *BZ* 102: 697–715.

Martin, T.H. 1854. *Recherches sur la vie et les ouvrages d'Héron d'Alexandrie, disciple de Ctésibus, et sur tous les ouvrages mathématiques grecs, conservés ou perdus, publiés ou inédits, qui ont été attribués à un auteur nommé Héron*, Paris.

Maspero, J. 1912. *L'organisation militaire de l'Egypte byzantine*, Paris.

Mathieu, M. 1961. 'Introduction'. In: *La Geste de Robert Guiscard*, Palermo: 3–96.

Mazzucchi, C.M. 1978. 'Dagli anni di Basilo Parakimomenos'. *Aevum* 52: 267–316.

McCabe, A. 2007. *A Byzantine Encyclopaedia of Horse Medicine, the Sources, Compilation and Transmission of the Hippiatrica*, New York.

McCormick, M. 1986. *Eternal Victory: Triumphal Rulership in Late Antiquity, Byzantium, and the Early Medieval West*, Cambridge.

McGeer, E. 1986. 'Menaulion-menaulatoi'. *Δίπτυχα* 4: 53–57.

McGeer, E. 1988. 'Infantry versus cavalry: The Byzantine response'. *REB* 46: 135–45.

McGeer, E. 1991. 'Tradition and reality in the "Taktika" of Nikephoros Ouranos'. *DOP* 45: 129–40.

McGeer, E. 1992. 'The Syntaxis armatorum quadrata: A tenth-century tactical blueprint'. *REB* 50: 219–29.

McGeer, E. 1995a. 'Commentary'. In: *Sowing the Dragon's Teeth: Byzantine Warfare in the Tenth Century*, Washington, DC: 171–360.

McGeer, E. 1995b. 'Byzantine siege warfare in theory and practice'. In: I.A. Corfis and M. Wolfe (eds.), *The Medieval City under Siege*, Woodbridge: 123–9.

McGeer, E. 2000. 'Commentary'. In: *The Land Legislation of the Macedonian Emperors*, Toronto.

McGeer, E. 2003. 'Two military orations of Constantine VII'. In: J.W. Nesbitt (ed.), *Byzantine Authors: Literary Activities and Preoccupations, Texts and Translations Dedicated to the Memory of Nicolas Oikonomides*, Leiden and Boston: 111–35.

McGrath, S. 1995. 'The battles of Dorostolon (971): Rhetoric and reality'. In: T.S. Miller and J. Nesbitt (eds.), *Peace and War in Byzantium: Essays in Honor of George T. Dennis*, Washington, DC: 152–64.

McGuckin, J.A. 2011–2012. 'A conflicted heritage: The Byzantine religious establishment of a war ethic'. *DOP* 65–66: 29–44.

McMahon, L. 2016. 'De Velitatione Bellica and Byzantine guerrilla warfare'. *Annual of Medieval Studies at CEU* 22: 22–33.

Mecella, L. 2009. 'Die Überlieferung der Kestoi des Julius Africanus in den byzantinischen Textsammlungen zur Militärtechnik'. In: M. Wallraff and L. Macella (eds.), *Die Kestoi des Julius Africanus und ihre Überlieferung*, Berlin: 85–144.

Meulder, M. 2003. 'Qui est le roi Mérops cité dans la Συλλογη Τακτικων?'. *Byzantion* 73: 445–66.

Mielczarek, M. 1993. *Cataphracti and Clibanarii: Studies on the Heavy Armoured Cavalry of the Ancient World*, Lodz.

Miller, T.S. 1976. 'The plague in John VI Cantacuzenus and Thucydides'. *GRBS* 17: 385–95.

Moffatt, A. and Tall, M. 2012. 'Introduction'. In: *Constantine Porphyrogennetos: The Book of Ceremonies*, Canberra: xxiii–xxxviii.

Moravcsik, G. 1938. 'Τα συγγράμματα Κωνσταντίνου του Πορφυρογεννήτου από γλωσσικής απόψεως'. *SBN* 5: 514–20.

Moravcsik, G. 1966. 'Klassizismus in der byzantinischen Geschichtsschreibung'. In: P. Wirth (ed.), *Polychronion: Festschrift für F. Dögler zum 75. Geburstag*, Heidelberg: 366–77.

Morris, R. 1988. 'The two faces of Nicephoros Phokas'. *BMGS* 12: 83–115.

Morris, R. 1994. 'Succession and usurpation: Politics and rhetoric in the late tenth century'. In: P. Magdalino (ed.), *New Constantines: The Rhythm of Imperial Renewal in Byzantium, 4th–13th Centuries*, Farnham and Burlington, VT: 199–214.

Müller, K.K. 1884. 'Neue Mittheilungen über Janos Laskaris & die Mediceische Bibliothek'. *Centralblatt für Bibliothekswesen* 1: 333–412.

Mullett, M. 1981. 'The classical tradition in the Byzantine letter'. In: M. Mullet and R. Scott (eds.), *Byzantium and the Classical Tradition, University of Birmingham Thirteenth Spring Symposium of Byzantine Studies 1979*, Birmingham: 75–93.

Mullett, M. 1997. *Theophylact of Ochrid: Reading the Letters of a Byzantine Archbishop*, Birmingham.

Negin, A.E. 1998. 'Sarmatian cataphracts as prototypes for Roman equites cataphractarii'. *Journal of Roman Military Equipment Studies* 9: 65–75.

Nelson, R.S. 2011–2012. 'And so, with the help of God: The Byzantine art of war in the tenth century'. *DOP* 65/66: 169–92.

Németh, A. 2010. *Imperial Systematization of the Past: Emperor Constantine VII and His Historical Excerpts*, PhD thesis, Central European University, Budapest.

Németh, A. 2013. 'The imperial systematization of the past in Constantinople'. In: J. König and G. Woolf (eds.), *Encyclopaedism from Antiquity to the Renaissance*, Cambridge: 232–58.

Neville, L. 2012. *Heroes and Romans in Twelfth-Century Byzantium: The Material History of Nikephoros Bryennios*, Cambridge.

Neville, L. 2016. *Anna Komnene: The Life and Work of a Medieval Historian*, Oxford.

Nicolle, D. 1980. 'The impact of the European couched lance on Muslim military tradition'. *The Journal of the Arms and Armour Society* 10: 6–40.

Nicolle, D. 1983. 'The Cappella Palatina ceiling and the Muslim military inheritance of Norman Sicily'. *Gladius* 16: 45–145.

Nilsson, I. 2006. 'To narrate the events of the past: On Byzantine historians, and historians on Byzantium'. In: J. Burke et al. (eds.), *Byzantine Narrative: Papers in Honour of Roger Scott*, Melbourne: 47–58.

Nilsson, I. 2010. 'The same story, but another: A reappraisal of literary imitation in Byzantium'. In: E. Schiffer and A. Rhoby (eds.), *Imitatio–Aemulatio–Variatio*, Vienna: 195–208.

Nilsson, I. and Scott, R. 2007. 'Towards a new history of Byzantine literature: The case of historiography'. *Classica et Mediaevalia* 58: 319–32.

Obolensky, D. 1971. 'Byzantine frontier zones and cultural Exchanges'. In: M. Berza and E. Stănescu (eds.), *Actes du XIVᵉ Congrès international des études byzantines*, vol. 1, Bucharest: 303–13.

Odorico, P. 1990. 'La cultura della Συλλογή. 1) il cosidetto enciclopedismo bizantino. 2) le travole del sapere di Giovanni Damasceno'. *BZ* 83: 1–21.

Odorico, P. 2006. 'Displaying la littérature Byzantine'. In: E. Jeffreys et al. (eds.), *Proceedings of the 21st International Congress of Byzantine Studies: London, 21–26 August 2006*, vol. 1, Aldershot and Burlington, VT: 213–34.

Oikonomides, N. 1971. 'L'organisation de la frontière orientale de Byzance aux Xᵉ–XIᵉ siècles et le *Taktikon de l'Escorial*'. In: M. Berza and E. Stănescu (eds.), *Actes du XIVᵉ Congrès international des études byzantines*, vol. 1, Bucharest: 285–302.

Oikonomides, N. 1972. 'Commentary'. In: *Les listes de préséance byzantines des IXᵉ et X siècles*, Paris: 281–363.

Oikonomides, N. 1988. 'Middle Byzantine provincial recruits: Salary and equipment'. In: J. Duffy and J. Peradotto (eds.), *Gonimos: Neoplatonic and Byzantine Studies Presented to Leendert G. Westerink at 75*, Buffalo, NY: 121–36.

Oikonomides, N. 1996. 'The social structure of the Byzantine countryside in the first half of the Xth century'. *Βυζαντινά Σύμμεικτα* 10: 105–25.

Ostrogorsky, G. 1953. 'Sur la date de la composition du Livre des Thèmes et sur l'époque de la constitution des premiers thèmes d'Asie Mineure'. *Byzantion* 23: 31–66.

Parani, G.M. 2003. *Reconstructing the Reality of Images (11th–15th Centuries)*, Leiden and Boston.

Peréz-Martín, I. 2008. 'El estilo Hodegos y su proyección en las escrituras constantinopolitanas'. *Segno e Testo* 6: 398–458.

Pertusi, A. 1948. 'Una acolouthia militare inedita del X secolo'. *Aevum* 22 (2/4): 145–68.

Pertusi, A. 1952. 'Commentary'. In: *Costantino Porfirogenito De thematibus*, Vatican City: 103–83.

Pertusi, A. 1956. 'Il preteso thema Bizantino di "Ṭālāja" (O Ṭājāla o Ṭāfālā) e la regione suburbana di Costantinopoli'. *BZ* 49: 85–95.

Petersen, L.I.R. 2013. *Siege Warfare and Military Organization in the Successor States (400–800 AD): Byzantium, the West and Islam*, Leiden and Boston.

Polemis, D. 1968. *The Doukai, a Contribution to Byzantine Prosopography*, London, Toronto and New York.

Pryor, J.H. and Jeffreys, E.M. 2006. *The Age of the ΔΡΟΜΩΝ: The Byzantine Navy ca 500–1204*, Leiden and Boston.

Ramadan, A.M.A. 2014–2015. 'The role of super natural powers in Arab-Byzantine warfare as reflected by popular imagination'. *Journal of Medieval and Islamic History* 9: 1–37.

Rance, P. 1994. *Tactics and Tactica in the Sixth Century: Tradition and Originality*, PhD thesis, University of St. Andrews, St. Andrews.

Rance, P. 2004a. 'The *Fulcum*, the Late Roman and Byzantine *testudo:* The germanization of Roman infantry tactics?'. *GRBS* 44: 265–326.

Rance, P. 2004b. 'Drungus, δρουγγος, and δρουγγιστί: A gallicism and continuity in Late Roman cavalry tactics'. *Phoenix* 58: 99–130.

Rance, P. 2005. 'Narses and the battle of Taginae (Busta Gallorum) 552: Procopius and sixth-century warfare'. *Historia* 54: 424–72.

Rance, P. 2007a. 'The date of the military compendium of Syrianus Magister (formerly the sixth-century Anonymus Byzantinus)'. *BZ* 100: 701–37.

Rance, P. 2007b. 'The Etymologicum Magnum and the fragment of Urbicius'. *GRBS* 47: 193–224.

Rance, P. 2008. 'Battle'. In: P. Sabin, H.V. Wees and M. Whitby (eds.), *The Cambridge History of Greek and Roman Warfare Volume II: Rome from the Late Republic to the Late Empire*, Cambridge: 342–78.

Rance, P. 2017a. 'The reception of Aineias' *Poliorketika* in Byzantine military literature'. In: M. Pretzler and N. Barley (eds.), *Brill's Companion to Aineias Tacticus*, Leiden and Boston: 290–374.

Rance, P. 2017b. 'Introduction'. In: P. Rance and N.V. Sekunda (eds.), *Greek Taktika: Ancient Military Writing and Its Heritage*, Gdańsk: 9–64.

Reinsch, D.R. 2010a. 'Der Autor ist tot–es lebe der Leser: Zur Neubewertung der Imitatio in der byzantinischen Geschichtsschreibung'. In: E. Schiffer and A. Rhoby (eds.), *Byzantium, Imitatio–Aemulatio–Váriatio*, Vienna: 23–32.

Reinsch, D.R. 2010b. 'Byzantinische Literatur–Tradition und Innovation'. In: V. Hirmer (ed.), *Byzanz: Pracht und Alltag; Kunst- und Ausstellungshalle der Bundesrepublik Deutschland, Bonn, 26. Februar bis 13. Juni 2010*, Munich: 56–61.

Riedel, M.L.D. 2010. *Fighting the Good Fight: The Taktika of Leo VI and Its Influence on Byzantine Cultural Identity*, PhD thesis, Exeter College, Oxford University, Oxford.

Riedel, M.L.D. 2012. 'Historical writing and warfare'. In: S. Foot and C.F. Robinson (eds.), *The Oxford History of Historical Writing Volume 2: 400–1400*, Oxford: 576–603.

Riedel, M.L.D. 2018. *Leo VI and the Transformation of Byzantine Christian Identity: Writings of an Unexpected Emperor*, Cambridge.

Roberto, U. 2009. 'Byzantine collections of Late Antique authors: Some remarks on the *Excerpta historica Constantiniana*'. In: M. Wallraff and L. Mecella (eds.), *Die Kestoi des Julius Africanus und ihre Überlieferung*, Berlin: 71–84.

Ross, M. 1958. 'Basil the proedros patron of the arts'. *Archaeology* 11: 271–5.

Roueché, C. 1988. 'Byzantine writers and readers: Storytelling in the eleventh century'. In: R. Beaton (ed.), *The Greek Novel AD 1–1985*, New York: 123–33.

Roueché, C. 2002. 'The literary background of Kekaumenos'. In: C. Holmes and J. Waring (eds.), *Literacy, Education and Manuscript Transmission in Byzantium and Beyond*, Leiden and Boston: 111–38.

Roueché, C. 2003. 'The rhetoric of Kekaumenos'. In: E. Jeffreys (ed.), *Rhetoric in Byzantium: Papers from the Thirty-Fifth Spring Symposium of Byzantine Studies, Exeter College, University of Oxford, March 2011*, Farnham and Burlington, VT: 23–37.

Runciman, S. 1929. *The Emperor Romanus Lecapenus and His Reign*, Cambridge.

Runciman, S. 1995. 'Historiography'. In: A.R. Littlewood (ed.), *Originality in Byzantine Literature, Art and Music: A Collection of Essays*, Oxford: 59–66.

Rydén, L. 1983. 'The "Life" of St. Basil the Younger and the date of the "Life" of St. Andreas Salos'. *Harvard Ukrainian Studies* 7: 568–86.

Saliou, C. 1996. 'Introduction and commentary'. In: *Le traité d'urbanisme de Julien d'Ascalon (VIᵉ siècle)*, Paris: 9–29, 79–132.

Sarraf, S. 2004. 'Mamluk *Furūsīyah* literature and its antecedents'. *Mamlūk Studies Review* 8: 141–200.

Savvides, A.G.C. 1993. 'Προσωπογραφικό σημείωμα για τον απελευθερωτή της μεγαλονήσου (965 μ.Χ.)'. *Επετηρίδα του Κέντρου Μελετών της Ιεράς Μονής Κύπρου Κύκκου* 2: 371–8.

Schindler, F. 1973. *Die Überlieferung der Stratagemata des Polyainos*, Vienna.

Schminck, A. 1986. *Studien zu mittelbyzantinischen Rechtsbüchern*, Frankfurt.

Schneider, R. 1908. *Griechische Poliorketiker*, vol. 2, Berlin.

Schreiner, P. 2011. 'Die enzyklopädische Idee in Byzanz'. In: P. Van Deun and C. Macé (eds.), *Encyclopedic Trends in Byzantium? Proceedings of the International Conference Held in Leuven, 6–8 May 2009*, Leuven, Paris and Walpole, MA: 3–25.

Scott, R. 1981. 'The classical tradition in Byzantine historiography'. In: M. Mullet and R. Scott (eds.), *Byzantium and the Classical Tradition, University of Birmingham Thirteenth Spring Symposium of Byzantine Studies 1979*, Birmingham: 61–74.

Searby, D.M. 2007. 'Introduction'. In: *The Corpus Parisinum: A Critical Edition of the Greek Text with Commentary and English Translation: A Medieval Anthology of Greek*

Texts from the Pre-Socratics to the Church Fathers 600 B.C.–700 A.D., 2 vols., Lewiston, NY: i.1–112.

Sekunda, N. 2008. 'Land forces'. In: P. Sabin, H.V. Wees and M. Whitby (eds.), *The Cambridge History of Greek and Roman Warfare Volume I: Greece, The Hellenistic World and the Rise of Rome*, Cambridge: 325–57.

Serikoff, N.I. 1992. 'Leo VI Arabus? A fragment of Arabic translation from the "Tactica" by Leo VI the Wise (886–912) in the Mamluk military manual by Iban Mankali (d. 1382)'. *Macedonian Studies* 9: 57–61.

Ševčenko, I. 1969–1970. 'Poems on the death of Leo VI and Constantine VII in the Madrid manuscript of Scylitzes'. *DOP* 23–24: 187–228.

Ševčenko, I. 1992. 'Re-reading Constantine Porphyrogenitus'. In: J. Shepard and S. Franklin (eds.), *Byzantine Diplomacy, Papers from the Twenty-Fourth Spring Symposium of Byzantine Studies, Cambridge, March 1990*, Aldershot and Burlington, VT: 167–96.

Shatzmiller, M. 1992. 'The crusades and Islamic warfare: A re-evaluation'. *Der Islam* 69: 247–88.

Shepard, J. 1985. 'Information, disinformation and delay in Byzantine diplomacy'. *BF* 10: 233–93.

Shepard, J. 1995. 'Imperial information and ignorance: A discrepancy'. *BS* 56: 107–16.

Shepard, J. 2001. 'Constantine VII, Caucasian openings and the road to Aleppo'. In: A. Eastmond (ed.), *Eastern Approached to Byzantium: Papers from the Thirty-Third Spring Symposium of Byzantine Studies, University of Warwick, Coventry, March 1999*, Aldershot and Burlington, VT: 19–40.

Shepard, J. 2002. 'Emperors and expansionism: From Rome to Middle Byzantium'. In: A. Abulafia and N. Berend (eds.), *Medieval Frontiers: Concepts and Practices*, Aldershot and Burlington, VT: 55–82.

Shepard, J. 2003a. 'The ruler as instructor, pastor and wise: Leo VI of Byzantium and Symeon of Bulgaria'. In: T. Reuter (ed.), *Alfred the Great, Papers from the Eleventh-Centenary Conferences*, Aldershot and Burlington, VT: 339–58.

Shepard, J. 2003b. 'The uses of "history" in Byzantine diplomacy: Observations and comparisons'. In: C. Dendrinos et al. (eds.), *Porphyrogenita: Essays on the History and Literature of Byzantium and the Latin East in Honour of Julian Chrysostomides*, Aldershot and Burlington, VT: 91–115.

Shepard, J. 2005. 'Past and future in Middle Byzantine diplomacy: Some preliminary observations'. In: M. Balard, É. Malamut and J.M. Spieser (eds.), *Byzance et le monde extérieur: Contacts, relations, échanges*, Paris: 171–91.

Shepard, J. 2012. 'Holy land, lost lands, *Realpolitik*: Imperial Byzantine thinking about Syria and Palestine in the later 10th and 11th centuries'. *Al-Qanṭara* 33: 505–45.

Sinclair, K.J. 2012. *War Writing in Middle Byzantine Historiography: Sources, Influences and Trends*, PhD Thesis, University of Birmingham, Birmingham.

Sinclair, K.J. 2014. 'Anna Komnene and her sources for military affairs in the *Alexiad*'. *Estudios bizantinos* 2: 143–85.

Smail, R.C. 1995. *Crusading Warfare (1097–1193)*, 2nd edition, Cambridge.

Sonderkamp, J.A.M. 1984. 'Theophanes Nonnus: Medicine in the circle of Constantine Porphyrogenitus'. *DOP* 38: 29–41.

Spanos, A. 2010. '"To every innovation, anathema" (?): Some preliminary thoughts on the study of Byzantine innovation'. In: H. Knudsen et al. (eds.), *Mysterion, strategike og kainotomia: Et festskrift til ære for Johny Holbek*, Oslo: 51–9.

Spanos, A. 2014. 'Was innovation unwanted in Byzantium?'. In: I. Nilsson and P. Stephenson (eds.), *Wanted: Byzantium: The Desire for a Lost Empire*, Uppsala: 43–56.

Speake, G. 1993. 'Janus Lascaris' visit to Mount Athos in 1491'. *GRBS* 34: 325–30.

Speidel, M.P. 1984. 'Cataphractarii clibanarii and the rise of the Later Roman mailed cavalry'. *Epigraphica Anatolica* 4: 151–6.

Spieser, J.M. 1973. 'Inventaires en vue d'un recueil des inscriptions historiques de Byzance'. *TM* 5: 145–80.

Stavridou-Zafraka, A. 1976. 'Ο ανώνυμος λόγος "Επί τη των Βουλγάρων Συμβάσει"'. *Βυζαντινά* 8: 344–406.

Stephenson, P. 2000. *Byzantium's Balkan Frontier: A Political Study of the Northern Balkans, 900–1204*, Cambridge.

Stephenson, P. 2010. 'The rise of the Middle Byzantine aristocracy and the decline of the imperial state'. In: P. Stephenson (ed.), *The Byzantine World*, London and New York: 22–33.

Stern, S.M. 1950. 'An embassy of the Byzantine Emperor to the Fatimid Caliph Al-Mu'izz'. *Byzantion* 20: 239–58.

Steward, M.E. 2016. *The Soldier's Life: Martial Virtues and Manly Romanitas in the Early Byzantine Empire*, Leeds.

Stouraitis, I. 2009. *Krieg und Frieden in der politischen und ideologischen Wahrnehmung in Byzanz (7.–11. Jahrhundert)*, Vienna.

Strässle, P.M. 2004. 'Krieg und Frieden in Byzanz'. *Byzantion* 74: 110–29.

Strässle, P.M. 2006. *Krieg und Kriegführung in Byzanz: Die Kriege Kaiser Basileios' II. gegen die Bulgaren (976–1019)*, Cologne.

Sullivan, D.F. 1997. 'Tenth century Byzantine offensive siege warfare: Instructional prescriptions and historical practice'. In: K. Tsiknakis (ed.), *Το Εμπόλεμο Βυζάντιο (9ος–12ος αι.)*, Athens: 179–200.

Sullivan, D.F. 2000. 'Introduction and commentary'. In: *Siegecraft, Two Tenth-Century Instructional Manuals by "Heron of Byzantium"*, Washington, DC: 1–23, 153–248.

Sullivan, D.F. 2003a. 'Introduction'. In: 'A Byzantine instructional manual on siege defense: The *De Obsidione Toleranda*. Introduction, English translation and annotations'. In: J.W. Nesbitt (ed.), *Byzantine Authors: Literary Activities and Preoccupations, Texts and Translations Dedicated to the Memory of Nicolas Oikonomides*, Leiden and Boston: 139–266.

Sullivan, D.F. 2003b. 'Byzantium besieged: Prescription and practice'. In: A. Abramea, A. Laiou and E. Chrysos (eds.), *Βυζαντινό κράτος και κοινωνία. Μνήμη Νίκου Οικονομίδη*, Athens: 509–21.

Sullivan, D.F. 2010a. 'Byzantine military manuals: Prescriptions, practice and pedagogy'. In: P. Stephenson (ed.), *The Byzantine World*, London and New York: 149–61.

Sullivan, D.F. 2010b. 'The authorship of Anna Komnene's *Alexiad*: The siege descriptions compared with the military instructional manuals and other historians'. In: A. Ödekan et al. (eds.), *First International Byzantine Studies Symposium: Change in the Byzantine World in the Twelfth and Thirteenth Centuries*, Istanbul: 51–6.

Sullivan, D.F., Talbot, A.-M. and McGrath, S. 2014. 'Introduction'. In: *The Life of St. Basil the Younger: Critical Edition and Annotated Translation of the Moscow Version*, Washington, DC: 1–61.

Syvänne, I. 2004. *The Age of Hippotoxotai: Art of War in Roman Military Revival and Disaster (491–636)*, Tampere.

Takirtakoglou, K.A. 2015. 'Οι πόλεμοι μεταξύ του Νικηφόρου Φωκά και των Αράβων'. *Βυζαντινά Σύμμεικτα* 25: 57–114.

Talbot, A.M. and Sullivan, D.F. 2005. 'Introduction'. In: *The History of Leo the Deacon, Byzantine Military Expansion in the Tenth Century*, Washington, DC: 1–52.

Tanner, R.G. 1997. 'The historical method of Constantine Porphyrogenitus'. *BF* 24: 125–40.

Theotokis, G. 2012. 'Border fury: The Muslim campaigning tactics in Asia Minor through the writings of the Byzantine military treatise *Περί παραδρομής του κυρού Νικηφόρου του βασιλέως*'. *ATINERS Conference Paper Series*, No: HIS2012-0308: 5–15.

Theotokis, G. 2013. 'The square fighting march of the crusaders at the battle of Ascalon (1099)'. *Journal of Medieval Military History* 11: 57–71.

Theotokis, G. 2014. 'From ancient Greece to Byzantium: Strategic innovation or continuity of military thinking?'. In: B. Kukjalko, O. Lāms and I. Rūmniece (eds.), *Antiquitas Viva 4: Studia Classica*, Riga: 106–18.

Tobias, N. 1979. 'The tactics and strategy of Alexius Comnenus at Calavrytae (1078)'. *Études byzantines* 6: 193–211.

Tobias, N. 2007. *Basil I Founder of the Macedonian Dynasty: A Study of the Political and Military History of the Byzantine Empire in the Ninth Century*, Lampeter.

Tougher, S. 1994. 'The wisdom of Leo VI'. In: P. Magdalino (ed.), *New Constantines: The Rhythm of Imperial Renewal in Byzantium, 4th–13th Centuries*, Aldershot and Burlington, VT: 171–80.

Tougher, S. 1997. *The Reign of Leo VI (886–912) Politics and People*, Leiden and Boston.

Toynbee, A.J. 1973. *Constantine Porphyrogenitus and His Work*, New York and Toronto.

Treadgold, W.T. 1979. 'The chronological accuracy of the *Chronicle* of Symeon the Logothete for the years 813–845'. *DOP* 33: 157–97.

Treadgold, W.T. 1992. 'The army in the works of Constantine Porphyrogenitus'. *RSBN* 29: 78–145.

Treadgold, W.T. 1995. *Byzantium and Its Army 284–1081*, Stanford, CA.

Treadgold, W.T. 2013. *The Middle Byzantine Historians*, New York.

Trombley, F.R. 1997. 'The *Taktika* of Nikephoros Ouranos and military encyclopaedism'. In: P. Binkley (ed.), *Pre-Modern Encyclopaedic Texts, Proceedings of the Second COMERS Congress, Groningen, 1–4 July 1996*, Leiden, New York and Cologne: 261–74.

Trombley, F.R. 2002. 'Military cadres and the battle during the reign of Heraclius'. In: G.J. Reinink and B.H. Stolte (eds.), *The Reign of Heraclius (610–641): Crisis and Confrontation*, Leuven, Paris, and Dudley, MA: 241–59.

Trombley, F.R. 2007. 'The operational methods of the Late Roman army in the Persian war of 572–591'. In: A.S. Lewin and P. Pellegrini (ed.), *The Late Roman Army in the Near East from Diocletian to the Arab Conquest: Proceeding of a Colloquium Held at Potenza, Acerenza and Matera, Italy (May 2005)*, Oxford: 321–56.

Tsagas, N.M. 1993. *Ιανός Λάσκαρις (1445–1535), Πρεσβευτής των Ελληνικών γραμμάτων στην Δυτική Ευρώπη*, Athens.

Tsamakda, V. 2002. *The Illustrated Chronicle of Ioannes Skylitzes in Madrid*, Leiden and Boston.

Tsougarakis, D. 1988. *Byzantine Crete, from the 5th Century to the Venetian Conquest*, Athens.

Tsurtsumia, M. 2011a. 'ΤΡΙΒΟΛΟΣ: A Byzantine landmine'. *Byzantion* 82: 415–21.

Tsurtsumia, M. 2011b. 'The evolution of splint armour in Georgia and Byzantium: Lamellar and scale armour in the 10th–12th centuries'. Βυζαντινά Σύμμεικτα 21: 65–99.

Van Bochove, T.E. 1996. *To Date and Not to Date: On the Date and Status of Byzantine Law Books*, Groningen.

Van den Berg, H. 1947. 'Introduction'. In: *Anonymus De obsidione Toleranda*, Leiden: 3–39.

Van Hoof, L. 2002. 'Among Christian emperors: The *Vita Basilii* by Constantine VII Porphyrogenitus'. *The Journal of Eastern Christian Studies* 54: 163–83.

Vannier, J.F. 1975. *Familles byzantines, les Argyroi: IXe–XIIe siècles*, Paris.

Vári, R. 1927. 'Die sogenannten Inedita Tactica Leonis'. *BZ* 27: 241–70.

Vasiliev, A.A. 1935–1968. *Byzance et les Arabes*, 3 vols., Brussels.

Vatchkova, V. 2011. 'La méthode byzantine de la *damnatio memoriae*'. In: A. Milanova, V. Vachkova and T. Stephanov (eds.), *Memory and Oblivion in Byzantium*, Sofia.

Vieillefond, J.R. 1932. 'Introduction'. In: *Jules Africain, fragments des Cestes provenant de la collection des tacticiens grecs*, Paris: vii–lviii.

Vieillefond, J.R. 1970. 'Introduction'. In: *Les Cestes de Julius Africanus: Étude sur l'ensemble des fragments avec édition, traduction et commentaries*, Florence and Paris: 5–70.

Vlysidou, B.N. 1985. 'Η συνωμοσία τοῦ Κουρκούα στο "Βίο Βασιλείου"'. Βυζαντινά Σύμμεικτα 6: 53–8.

Vlysidou, B.N., Kountoura-Galaki, E., Lambakis, S., Lounghis, T. and Savvidis, A.G. 1998. Η Μικρά Ασία των θεμάτων: έρευνες πάνω στην γεωγραφική φυσιογνωμία και προσωπογραφία των Βυζαντινών θεμάτων της Μικράς Ασίας (7ος–11ος αι.), Athens.

Vogel, E.G. 1854. 'Litterarische Ausbeute von Janus Lascaris, Reisen im Peloponnes um's Jahr 1490'. *Serapeum* 15: 154–60.

Von Sievers, P. 1979. 'Military, merchants and nomads: The social evolution of the Syrian cities and countryside during the classical period, 780–969/164–358'. *Der Islam* 56: 212–44. Vratimos, A. 2012. 'Was Michael Attaleiates present at the battle of Mantzikert?'. *BZ* 105: 829–40.

Vryonis, S.J. 1957. 'The will of a provincial magnate, Eustathius Boilas (1059)'. *DOP* 11: 263–77.

Walker, P.E. 1977. 'The "crusade" of John Tzimisces in the light of new Arabic evidence'. *Byzantion* 47: 301–27.

Wallraff, M., Scardino, C., Mecella, L. and Guignard, C.J.-D. 2012. 'Introduction'. In: *Iulius Africanus Cesti, The Extant Fragments*, trans. W. Adler, Berlin: xi–xcii.

Webb, R. 2009. *Ekphrasis, Imagination and Persuasion in Ancient Rhetorical Theory and Practise*, Farnham and Burlington, VT.

Whately, C. 2015. 'The genre and purpose of military manuals in Late Antiquity'. In: G. Greatrex, H. Elton and L. McMahon (eds.), *Shifting Genres in Late Antiquity*, Farnham and Burlington, VT: 249–61.

Wheeler, E.L. 1988. *Stratagem and the Vocabulary of Military Trickery*, Leiden.

Wheeler, E.L. 2013. 'Polyaenus: *Scriptor Militaris*'. In: K. Brodersen (ed.), *Polyainos: Neue Studien/Polyaenus: New Studies*, Berlin: 7–54.

Whitby, M. 1988. *The Emperor Maurice and His Historian: Theophylact Simocatta on Persian and Balkan Warfare*, Oxford.

Whittow, M. 1996. *The Making of Orthodox Byzantium 600–1025*, Berkeley, CA and Los Angeles.

Wojnowski, M. 2005. 'Κατάφρακτοι–Ciężkozbrojna jazda Cesarstwa Bizantyjskiego jako kontynuacja antycznych Cataphracti I Clibanarii'. *Zeszyty Naukowe Uniwersytetu Jagiellońskiego* 21: 8–22.

Wojnowski, M. 2012. 'Periodic revival or continuation of the ancient military tradition? Another look at the question of the kataphraktoi in the Byzantine army'. *Studia Ceranea* 2: 195–220.

Zuckerman, C. 1990. 'The military compendium of Syrianus Magister'. *JÖB* 40: 209–24.

Zuckerman, C. 1994. 'Chapitres peu connus de l'Apparatus Bellicus'. *TM* 12: 359–89.

Zuckerman, C. 2014. 'Squabbling *protospatharioi* and other administrative issues from the first half of the tenth century'. *REB* 72: 193–233.

Index

3 20